T0093715

HOW TO FIGHT A WAR

HOW TO FIGHT A WAR

***

Mike Martin

HURST & COMPANY, LONDON

First published in 2023
in the United Kingdom by
C. Hurst & Co. (Publishers) Ltd.,
New Wing, Somerset House, Strand,
London WC2R 1LA
www.hurstpublishers.com

ISBN 9781787389304

Distributed in the United States, Canada and Latin America
by Oxford University Press, 198 Madison Avenue,
New York, NY 10016, United States of America.

A Cataloguing-in-Publication
data record for this book is available
from the British Library.

Cover design and line illustrations,
Lyn Davies Design. Text design and setting,
Sathi and Prerana Patel and Lyn Davies.
Maps by Sebastian Ballard.
Printed in India

For Sarah and Elsie

CONTENTS

Advance Praise for 'How to Fight a War'

'Colourful, punchy, admirably challenging and clear–essential reading for every soldier, officer and General.'

General Sir Patrick Sanders (KCB, CBE, DSO, ADC Gen), *Chief of the General Staff*, British Army (2022–)

'At a time of high geopolitical worry and risk this book sets out clearly the complex considerations–too often insufficiently assessed–which ought to inform any decision, by anyone, about going to war.'

The Rt Hon. Charles Clarke, Co-Lead, *University of Cambridge Baltic Geopolitics Programme*, former MP, Home Secretary (2004–06)

Rt Hon Johnny Mercer MP, *Minister for Veterans' Affairs*

'War is abhorrent and complicated. But it is also human. In this easy-to-read book Mike Martin strips away much of its technical characteristics and reminds us that war is determined by some basic tenets that only a fool ignores. How to Fight a War is for statesmen, diplomats and generals to keep by their side when conflict beckons.'

Lieutenant-General Douglas Chalmers (CB, DSO, OBE), *Deputy Chief of the Defence Staff, British Army 2018–2021, and Master of Emmanuel College, Cambridge University*

'The ongoing war of Russian aggression in Ukraine has caught the majority of Western policymakers and the public off guard and laid bare a severe deficit in understanding the fundamentals of modern warfighting and military affairs. This new book will help address this dearth of expertise on military matters in the West by succinctly summarizing the basic principles of war and warfare in layman's terms.'

Franz-Stefan Gady, *Senior Fellow, IISS*

'Almost all of the wars that have been fought over the last thirty years have been unsuccessful. If the leaders, generals and statesmen had read this book, they might not have been.'

'An outstanding, clearly written and articulated book for both undergraduates and graduates studying strategic or defence studies, not to mention junior to mid-ranking officers, even senior ones.'

Ahmed S. Hashim, *Associate Professor of Strategic Studies, Deakin University and Australian Defence College,* author of *Insurgency and Counter-Insurgency in Iraq*

List of Figures and Maps

Introduction:

Why You Should Read 'How to Fight a War'

Misunderstandings abound about what war is, and what it isn't. This is true not only for civilians and the public, but also for generals and political leaders—those whose responsibility it is to think about wars, and particularly how to win them. A brief glance at the record of the nineteenth and twentieth centuries shows that far more wars are lost, or stumble towards an inconclusive draw, than are won. What is going wrong? Why do so many leaders make catastrophic mistakes and lead their militaries and countries to defeat?

At the core of this book is the notion that winning wars is about understanding and following basic principles. Although war is fiendishly complex, wars are almost always lost due to the same simple ideas being misapplied or ignored. War is a psychological phenomenon—a competition between evolved human brains—and its core nature has altered hardly at all since humans first began to live in groups hundreds of thousands of years ago.

At first glance, the notion that war hasn't changed seems wrong. Humans began fighting wars with clubs, flint axes and throwing rocks. We now fight wars with hypersonic missiles and cyber-attacks. The technology with which wars are prosecuted has changed (and is changing) beyond all recognition. But the point of all this technology—indeed of the entire physical apparatus of war—is primarily as a tool to affect the psychology of your opponents, and especially of the opposing leader or commander. Ultimately war is about changing their minds, and making them see the world differently, or dying.

The dynamics of war—advance, retreat, flank, bluff, rout, deceive, encircle—occur in every conflict, and at every level. This is why the practitioners of today's wars study previous battles. Strip away the technology, and the importance of strategy and intelligence then, and now, are equal. It is also why the only way to become a general in charge of a division is to first command platoons, companies, battalions, and brigades. The dynamics are the same, albeit at an ever-larger scale.

The idea that war is timeless, which derives from its foundations in human psychology, is a big one. It is the dominant concept in the study of war, from which all else flows, and is a theme that recurs throughout this book. It also provides the foundations for the hard elegant logic of war: the explanation of why things are the way that they are in warfare.

When war leaders fail in their aims, it is usually because they have ignored warfare's simple ideas, thinking that, for instance, logistics matter less to them than to their adversaries. This 'wishing away' happens because of three fallacies: overconfidence; being bewitched by a new technology that will 'solve' their problems; or misunderstanding the enemy and their perspective. And these three fallacies—rooted as they are in human psychology—recur again and again in human history and politics.

After a long period of the declining frequency of war, we are likely to be fighting more wars in the coming years as the post-Second World War consensus breaks down, and global challenges like climate change, migration, and the technological revolution proceed apace and remain unaddressed. At the time of writing, we are amid a war in Europe, the first since 1945, which has four of the five main nuclear powers involved. Geopolitical tension in the Far East, not least over China, Taiwan and the South China Sea is here to stay. The Middle East and the Sahel region of Africa are highly unstable, and becoming more so. The world has not looked this chaotic for decades.

* * *

How to Fight a War was written as a reference guide for the Commander in Chief of a nation's military. In an age of inevitable and more frequent wars, our leaders must have the strategic, operational, and tactical skills to prosecute wars successfully. The ability to do so means that we may arrive at durable strategic answers to the pressing geopolitical questions of the day quicker and more efficiently than might otherwise be the case. This makes me sound like a warmonger, which I most certainly am not, having experienced war first-hand.

Successfully prosecuting a war is far better than fighting one that drags on without delivering its intended geopolitical outcomes, with the concomitant needless wasting of soldiers' and civilians' lives and widespread death and destruction.

* * *

Are there any overall insights about war that you, the leader, must grasp beyond its rooting in human psychology?

The first key lesson is that war is political. In a famous encapsulation, war is simply politics by other means. Very often you will see war on the one hand, and politics and diplomacy on the other, discussed as if they were discrete spheres of activity, with only the narrowest of connections between them.

War is a subset of politics. It is how humans conduct politics when they have failed to reach a decision through talking. It is, some have called it, armed politics.

If war is politics by other means, if it is what we do when we have run out of things to say, then the violence in war is a method of communication. At first glance, this seems abhorrent, or crazy. Indeed, when discussing war with people with no military background, they think that I have lost the plot. But it is a logical deduction from the underlying principle that war is a political act. If you do not believe me now, you may do so by the time you get to the end of the book.

Violence as communication functions in a tactical manner. If there are enemy troops occupying a hilltop that you wish to seize, you can ask them to leave. If you are in a fighting war and those soldiers are well supplied and maintain high morale, they are likely to demur. What other mechanisms do you have at your disposal to underscore the point that you really want them to vacate the hilltop? You will soon resort to lethal violence where you set out to kill the enemy. In so doing, you are saying, I really want that hilltop, and if you don't surrender, or abandon the position, you will be killed. Lethal violence is simply communication that your adversary cannot ignore.

Violence as communication is also true in a deeper strategic sense. We can see this when a state launches a single attack—a missile or an air strike on a chemical weapons research facility—at another country. The likelihood is that this attack would not happen in isolation. There would have been a long series of discussions and statements between the countries trying to convince the attacked country to do something, or not do something—in this example not to continue its chemical weapons research. Here the missile attack is used to underscore the point—we really want you to stop chemical weapons research—rather than as a way of halting production.

And if war is politics by other means, then the distinctions of war versus peace are not as sharp as we might think. Very often in popular discourse war and peace are presented as binary opposites, as different states of being, with one being inherently bad, and the other inherently good. But peace is not the absence of war. Peace is the ability to handle conflict by peaceful means. It is the building of human political structures that enable us to keep talking, so that we need not resort to lethal violence to communicate.

War and peace are a continuum of activities seeking a conclusion to a geopolitical question. Out of that resolution will come new human political structures, which will encapsulate and channel future urges towards violence into talking. Sometimes we have to fight a war to reach a durable peace; sometimes imposing peace onto a war before its questions are settled simply sets the scene for the next round of hostilities.

Because war and peace are a continuum, your war strategy starts when you are still at peace. Moreover your strategy for war must be to avoid war at all costs because it is so ruinous to life, society and property. You never get a dead son or daughter back, and your village will never be rebuilt in the way it was previously. One of the best ways to avoid war therefore is to deter your enemies, and to signal to them that you mean business, while minding your own business.

Paradoxically, you may decide that the way to not go to war is to retain a large and capable military, because it makes potential adversaries think twice. Deterrence is a central plank in strategies to avoid war and you should pay great attention to it as leader. Maintaining a military with full spectrum capability and global reach is extraordinarily expensive. If you cannot afford this—and at present only the United States can—then you must either reduce the number of capabilities you wish to field, or restrict your area of strategic ambition to a sub-global region. Weak militaries that only look good on paper will get humbled along with the country that declined to resource them properly.

There is one final insight that you should consider as you lead your country in war. War is not a rational act. There is reasoning involved in it for sure—you as a leader must think logically and clearly to construct and conduct your strategy. Without this, you will not win. But if you think about it, war is completely illogical. Why would an individual put their life on the line, or risk disfigurement and injury, to advance the positions of their compatriots, or co-tribals, or co-religionists? In the defensive register, war makes more sense—you are defending your family and your home—but in the offensive, the risk-reward calculus defies logic.

While war is a psychological act, humans are not rational beings; rather, they are emotional beings. Hence the guiding structures of war are emotional—concerning pride, and belonging, and status, and jealousy, and fear—about which we try to think logically in order to prevail. War is both an art and a science.

The big mistake that many make is to see war as a rational, logical act—conflict by spreadsheet—when it is anything but. And this means that intangibles, such as morale and strategy, are far more important than the amount or type of equipment or technology that you have. This key tenet is reflected in the chapter order of *How to Fight a War*, with the most important ones—those that have a greater bearing on the chances of victory—coming first.

* * *

Chapter 1 looks at the art of strategy, and what enables it: an understanding of your enemy, and of the world. Strategy is supreme above all else in warfare, and it is the greatest of the intangible factors of war that make up Part 1 of this book. This chapter tells you how to think about strategy, how to understand your (human) weaknesses as you try and form strategy, and how to design your organisations so that they deliver you good intelligence and help you form good strategies.

In the words of an old sage of warfare: strategy without tactics is the slowest route to victory, but tactics without strategy is the noise before the defeat. If you only get one thing right as a leader, opt for a correct strategy before anything else.

Chapter 2 explores logistics. It is so much more than getting materiel to where it is needed, more even than a huge system spanning continents, but rather an underpinning philosophy in how you fight your wars: you must always do so in a way that protects your logistics and undermines or destroys your enemy's. Again the comparison with tactics, but another well-worn maxim in the command structure of successful armies is 'amateurs talk tactics, professionals talk logistics'. As a leader, once you have worked on the strategy, you must always base your plans on sound logistics.

Next, in Chapter 3, you will read about morale, how to build it, and how to sustain it. You will learn that morale is like a glue that holds your forces together and inoculates them against fear. Forces that remain cohesive and confident tend to win battles over those that are fragmented and fearful. And so in the same way that you seek to protect your logistics, and destroy your enemy's, your goal is to bolster your morale and undermine your enemy's. As an old general said: morale is to the physical, as three is to one.

Last, in Chapter 4 we turn to training. This is how you prepare your people and create your teams for specific tasks amid the extraordinary complexity of modern warfare. Training creates physical and mental toughness to survive the 24-hour-a-day, 365-days-a-year nature of combat. Better trained people are more likely to have resilient morale and more likely to survive combat. And when your people are all trained to the same standards it makes it easy to reorganise your army in the middle of a war, detaching this unit to that unit, and to supply them through the same logistical system. Common training makes them feel like an army rather than a series of factions fighting on the same side.

Together these four elements—strategy (and intelligence), logistics, morale, and training—form an intangible foundation on which your military fights. The critical point here is that the intangibles of war are far more important than technology, or even the size of the armed force that you are fighting with or against. It is a common misconception to focus on what you can count—the number of tanks, for instance—but if they are not deployed under the right strategy, with enough fuel and ammunition to fight, and with crews that are of high morale and well trained, then those tanks become expensive targets.

* * *

In Part 2 you will learn about the different domains of warfare. You will discover that the land environment is the most important domain in which humans fight—what is known in military parlance as the decisive domain—where geopolitical questions are decided. This is because, put simply, that is where humans live, and wars will ultimately be decided when one side has its infantry in the other side's cities and villages. This chapter will also teach you about what makes up a land army, and how all of its various sub-components—such as infantry and artillery and engineers—work together.

Chapter 6 discusses the sea, air and space forces who operate in support of those on land. You will learn that the maritime domain is primarily about protecting your global logistics (and denying your enemy the ability to do the same), and secondarily about launching weapons against targets ashore. Air forces operate mainly in direct support of land forces either transporting them, reconnoitring their environment, or delivering ordinance where it is required (and denying your enemy the same). Space forces, a rapidly developing area of military science, offer extraordinary opportunities for communications and reconnaissance.

Chapter 7 will take you through the information and cyber operations that operate in support of land forces. Much heralded in recent years as the latest new way of war, the chapter sets out the realistic limits of what can be achieved with ones and zeros. It is an important subject, and a rapidly developing one, but is far from achieving a decisive effect on the geopolitical environment on its own. Chapter 8 will delve into nuclear, biological and chemical weapons and whether they will, or will not, be used in the coming wars. Taken together, the chapters in Part 2 will help you understand the vast range of capabilities and technology at your disposal as a Commander in Chief.

* * *

Part 3 brings it all together and shows you how to orchestrate lethal violence to achieve your political goals. In other words, how to change your enemy's mind, or kill them. It describes the fog of war—the friction and the ambiguity—in which you must make quicker decisions than your enemy.

It describes *How to Fight a War*.

Part 1

INTANGIBLE FUNDAMENTALS

1.

STRATEGY AND INTELLIGENCE

The single most important thing that the leader of any military force must do is to develop a realistic strategy. That this is salient in the practice of warfare has been known for millennia. Indeed, Sun Tzu wrote 2,500 years ago that 'strategy without tactics is the slowest route to victory; [but] tactics without strategy is the noise before the defeat'.

Having an unrealistic or otherwise flawed strategy is the most common mistake that leaders make when committing their forces to war. Your country, empire or coalition may have the biggest, best-equipped and most highly trained army in the world, but without a realistic strategy set by the leader, your war will always fail in its aims. So, although what follows is one of the shorter chapters in *How to Fight a War*, if you only read one chapter, read this one.

Originally from the Greek word *strategia* meaning generalship, 'strategy' is one of the most misused words in the English language. Often it is confused with 'plan', as in a sequence of actions, whereas a 'strategy' is far more all-encompassing. It comprises an understanding of the world, a set of overarching, high-level objectives or goals, a description of the methods to achieve those goals (a plan), and which resources are needed and should be used. Generalship is much more than just a plan.

You can tell when a country or a leader lacks a realistic strategy because they tend to list either activities ('we are launching airstrikes') or ill-defined goals ('X country must lose') in the place of an unambiguous, realistic goal ('we are going to remove the armed forces of X country from the territory of Y country'). In other words, they are confusing activity for output.

Another obvious 'tell' is when a country's war aims keep shifting from one thing to another: this reveals that they have not thought through their overarching objectives and are being blown from one to the other as events unfold. A leader and a country must carefully select, and then continue to follow, their overall strategic objectives: this is a fundamental principle of waging war.

A good strategy should contain clear, simple objectives. In the Second World War, the Allied Powers' (the US, Britain, the Soviet Union, and others) goal was the total defeat and unconditional surrender of the Axis Powers (Germany, Italy, Japan, and others). These defeats were to be sequenced: the Allies decided to ensure the defeat of Germany and Italy before the defeat of Japan. While Germany and Italy were being defeated, the Allies fought a lower prioritised war against Japan. This was because in the realistic intelligence assessment of the Allies, Nazi Germany represented a much stronger and more dangerous foe who, were they able to defeat the Russians and the British in Europe, might become unassailable.

This was known as the 'Europe First' plan. In a testament to the enduring nature of the Allies' strategy, it survived even the Japanese attack on the United States at Pearl Harbor, where quite reasonably the United States might have reconsidered that Japan was the more immediate threat. The resources required for this strategy precipitated a series of other objectives, not least among them winning the Battle of the Atlantic, so that Allied supplies could cross from the United States to Britain and Russia. And it was because of the immense resources required that the Allies decided in the first place to sequence the defeat of their enemies, rather than to take them on simultaneously.

You will see from this example that the components of strategy can be shortened to Ends (overarching objectives: the defeat of the Axis powers), Ways (the sequence of actions, or plan: the Europe First plan) and Means (the resources required to carry out your plan: American supplies used by Britain and Russia). Your strategy—with its Ends, Ways, and Means—must rest on a solid, objective understanding of the problem you are trying to solve with military force, and of the wider world in general (the Allied appreciation of the relative strengths of Germany and Japan).

A good strategy should also offer you a framework against which to judge potential actions. That is, if you do X or Y, which of the two is more likely to achieve your eventual aim(s)? This is particularly important in war, because for most of the time the ongoing fighting is at best confusing. It will be hard to discern who is winning a particular battle, or (especially) what the enemy leader is thinking. A realistic strategy will help you cut through that ambiguity and focus your limited resources on achieving your overriding objectives.

In early 1941 the British Prime Minister Winston Churchill was faced with the critical decision of whether to keep several hundred newly-produced tanks in Britain to help defend it from invasion, or to send them to the Middle East where they would help defend Egypt (at that time under British control) from the German and Italian armies.

In the end the decision was made to deploy the tanks to the Middle East where they would help secure British oil production and the strategic prize of the Suez Canal which at that time supplied Allied forces in the Mediterranean, the Middle East, and the Far East. It was judged that were the Canal and the oil to fall into enemy hands, Britain would likely lose the war. There was no option but to send the tanks to Egypt, even while incurring the risk of Britain being less prepared to defend itself in the event of invasion.

The best way to visualise strategy is as a three-legged stool with Ways, Ends and Means each comprising one leg, and with the stool resting on the foundations of a sound understanding of the problem/enemy you face, and the worldwide context (see Fig. 1).

In theory it is simple to decide upon a strategy. But in practice it is rarely achieved. This is because the simplicity of the concepts outlined above are often obscured by humans' inability to understand the world objectively, or think through strategy clearly, without falling foul of their own cognitive biases. When leaders or countries produce poor strategies, it is because they have been unable to escape their own biases. These biases exist because of the way that human cognition evolved, and are completely unavoidable. If you think that you do not have cognitive biases, you should beaware that one of the main cognitive biases is thinking that you do not have any biases.

How cognitive biases affect intelligence understanding and strategy formation

Understanding the world using intelligence and forming strategy are both prone to several well-established human cognitive biases. The most common of these is overconfidence (or hubris), something that leaders are often prone to, and which you must guard against. The ease with which British and American leaders felt that they would reconstitute Iraq after the 2003 invasion is one obvious recent example, which meant that they did little or no planning for the aftermath of the war, and the country quickly descended into chaos.

This overconfidence is exacerbated by obsequiousness bias, namely our tendency to agree with the statements of those above us in the hierarchy. If you the leader projects confidence (as leaders do), your troops will often nod along, not wishing to go against the hierarchy. The Coalition and then NATO operations in Afghanistan, from 2001 to 2021, offer a salutary tale: every year the officers in the field reported steady progress, and that the coming year would be the decisive one, where a corner would finally be turned. In reality—and as evidenced by the collapse of the Afghan government and security forces before NATO had even left—there was little or no progress being made. As leader, these two biases are the hardest for you to counteract—resting, as they do, upon your own personal psyche and power base. You must continually guard against them.

The other type of bias that you must beware of—and this is linked to overconfidence—is becoming sucked into a status dispute with your opposing leader. That is, of personalising the war and making it about You versus the Enemy Leader. All leaders are driven by achieving status and have become leaders by seeing off multiple status challenges in their own political parties, militaries or countries. Once they become leader of their group, this habit of seeking, guarding against, and winning status challenges transfers to the leaders of other groups. And when this happens, wars can become about beating the other leader, or humiliating them, rather than achieving strategic outcomes for your country.

The other important category of biases that affect intelligence analysis and strategy formation are those that complicate how we understand in-groups and out-groups. Humans have a natural inclination to categorise between the two: the in-group is Us, our family, tribe or country, and the out-group is Them, the other family, tribe or foreign country. We evolved this cognitive mechanism for sound reasons: when we lived in small bands on the African savannah over 100,000 years ago, it enabled us to grow social groups that were bigger than those related by blood, because it allowed us to categorise who 'should' and 'should not' be members of such groups.

We treat in-groups and out-groups differently. In-groups are trusted, assumed to be guided by positive motives and honest. We also assume that the way things are done by our in-group is the norm for all humans (our culture, or our political system, for instance). On the other hand, out-groups are often distrusted, thought dishonest or have their motives questioned or considered negatively. The way they do things is considered abnormal. If you believe that all groups have the same ratios of honest to dishonest, and positive to negative, people among them, then you will appreciate how this bias creates an out-group blind spot which hampers strategy formation.

This happens in two ways. First, misunderstanding the motives of your opponent or enemy will cause you to misjudge the amount of support that they have both within their country and beyond. Thinking that they are dishonest, when their entire population considers them to be honest, for instance, would be a warning sign worth heeding. You will also likely misunderstand the morale and cohesion of their armed forces, which are fundamental to how effective they will be (see Chapter 3). These misjudgements about the other are also exacerbated by overconfidence and obsequiousness biases.

Second, the out-group blind spot causes leaders to assume that the way they see things—for example, that a particular territory has always been their ancient homeland, or that a form of government is unrivalled—is the normal way of seeing the world, and all others are abnormal and must be changed. This is also known as the mirroring bias. Hence leaders that go to war tend to frame strategic problems incorrectly, often setting objectives unconnected with reality, but mirroring *their* reality.

The 2022 Russian invasion of Ukraine offers a good example of this defect. Vladimir Putin thought of Ukraine as part of Russia, whereas it was and is an independent country. He also understood the government to be run by 'Nazis' who were oppressing the population whereas, although imperfect, the Ukrainian government had wide support before and especially after the Russian attack, leading to determined resistance against the invaders. As the Russians sought to understand the Ukrainian environment, they were clearly beset by confirmation bias, with each piece of new intelligence gathered confirming their already held (incorrect) assumptions.

Another example of the mis-framing of a strategic problem is that of the post-2001 Global War on Terror, where multiple internal conflicts in various Muslim countries—Somalia, Yemen, Afghanistan, Iraq, etc.—were shoehorned into an overarching narrative of fighting global terrorism. Many of these disputes had arisen for different reasons—poor governments, internal ethnic or tribal divides, or resource inequality between groups—and their erroneous framing exacerbated many of these conflicts. The same thing happened during the Cold War when multiple internal conflicts were seen through the global narrative of capitalism versus communism.

These biases can be surmounted. Indeed, human cognitive biases must be overcome if you wish to create realistic and achievable strategies. And the way that you overcome them is to design the organisations that surround you as a leader in a way that minimises these unavoidable biases by forcing individuals, teams, and ideas, to compete with, and question, each other. These organisations and processes should be split into two: those that contribute to your understanding of the enemy and the wider world, and those that help you form strategy. Each will now be looked at in turn.

How to understand your enemy, and the world, clearly

Information about the world will come to you in two forms. First, there is the public understanding that you can read in the *New York Times* or on the internet. Second, as the leader of a country, you will have access to intelligence networks, organisations and capabilities. Intelligence is information that has been brought together to answer questions. It derives from many sources, including secret ones. The detailed exposition of the different methods of collecting intelligence is beyond the scope of *How to Fight a War*. But it is critically important for you, as leader, to understand the advantages and disadvantages of each collection method (and of intelligence in general), and how raw intelligence should be analysed so that sound conclusions are drawn while avoiding cognitive biases.

As a guiding principle, your intelligence operation should be telling you about the intentions, motivations and capabilities of your enemy. This can be summarised as the What (What are their plans?), Why (Why do they have those plans/what is their world view?), and How (How do they intend to achieve this—do they have a new weapons system, for instance?).

You must keep this entire operation, and the intelligence that it generates, completely secret, not only because you must protect your sources and methods of collection from compromise, but also because once you have gained specific information you must ensure that your enemy doesn't know what you know.

At this level of understanding, you are trying to build a strategic picture of the world, and so your enemy's intentions and motivations should be accorded more importance than their capabilities. At the tactical level (discussed in Chapter 9), information on the capabilities—that is, the troops, weapon systems and supplies—of your enemy attain paramount importance. (It is moreover more difficult to assess intentions and motivations in the fluid environment of the battlefield where bluffs and ruses abound).

You should situate this specific, secret knowledge about an enemy or opponent in a wider world view that is gained not from secret intelligence but from your diplomats, the press and academia. How does your enemy's economy mesh into the regional and global economy? Who are their allies at the United Nations? What are the common narratives about their leader in the world press? Do other world leaders respect and/or like him or her?

Your intelligence agencies should routinely gather open-source information. Historically this involved going to trade fairs, but today it means gathering information from the internet, and particularly social media.

The main types of collecting secret intelligence are by talking to people (human intelligence), intercepting their communications (signals intelligence), taking pictures from aircraft or satellites (image intelligence), and measuring emissions and signatures of equipment (electronic intelligence).

Human intelligence—commonly known as spying—has one key advantage above all the other types of intelligence: your agents can access the inner thoughts and motivations of your intelligence targets in a way that other forms of intelligence rarely can. You will, for instance, find out about leadership dynamics in the enemy's inner circle. But this comes with two major caveats that you must always bear in mind: first, human intelligence is, by definition, one person's view and subjective; and second, your source may realise that they are being used as a source and be deliberately influencing your views.

Signals intelligence is the interception of communications. Traditionally this would rely upon intercepting your enemy's letters and battlefield dispatches. Nowadays this means intercepting internet communications, and particularly those on smartphones (due to the diverse and large volumes of data that we all keep on our smartphones), as well as gaining access to your enemy's encrypted military and government communications. The most powerful and successful signals intelligence organisation in history is the 'Five Eyes' network comprising the US, the UK, Canada, Australia, and New Zealand. My strong advice would be not to fight a war against any of the Five, or if you find yourself doing so, to rigorously, forensically and continuously assess and reassess the security of your communications.

Signals intelligence can give you specific insights into an intelligence target's planning, for instance their travel itineraries, enabling you to strike them with a drone, but is less good at providing information on their motivations or inner thoughts. Signals intelligence is also context free—the intelligence product will only convey what a particular person said at a particular time. This means that human intelligence and signals intelligence can be very powerful when paired together. You should always remain wary of an assessment that is drawn solely from human intelligence or solely from signals intelligence.

Image intelligence is the analysis of images taken from aircraft or satellites. Nowadays these can be of sufficient resolution to enable you to tell if a particular person has arrived for a meeting, and will certainly be good enough to enable you to map out the dispositions of your enemy's forces and the types of capabilities deployed on the battlefield (notwithstanding any camouflage or deception that they may employ).

This can often be successfully coupled with electronic intelligence: the measurement of emissions from equipment like radars, power plants, vehicles and aircraft. It can be incredibly powerful in giving a detailed picture of the equipment and capabilities that your enemy has deployed: some countries can determine which type of aircraft is flying based upon the turbulence signature it generates. There are other types of intelligence, but these are the four main types used in military affairs.

Each of these methods of intelligence collection will allow you access to information that your enemy doesn't want you to know.

Equally as useful, it will also give access to information that they don't know that you know. Both are very useful and will help to inform your strategy and plans.

However, you must remain alert to the two main shortcomings of all forms of intelligence: first, it is not, and can never be, the whole story. Your intelligence picture is likely to be a composite one made up of individual pinpricks of information. And second, there is a natural tendency for leaders and officials to ascribe more weight to intelligence that is 'secret' or 'clandestine' (essentially because it is glamorous). Hence secret information is considered more 'correct' than open-source information you can get from reading *The Economist*.

As leader, you must ensure that your intelligence organisations guard against being undermined by such errors, such as analysing several context-free pinpricks of secret information—the travel plans of their leadership gained from hacking their mobile phones, along with a human source who reveals a plan to launch an attack next week—and using them as the sole basis for understanding. In this imagined scenario there might be extensive open-source reporting describing the strategic situation as unfavourable to an attack next week. To consider secret intelligence without enmeshing it within such wider contextualisation may result in the wrong conclusions being drawn.

For NATO forces in Afghanistan this was a recurring problem, because in a society with low levels of literacy there was little open-source information available on contextually important matters: tribal structures, religious networks, and land ownership. NATO often relied on an intelligence picture heavily skewed towards secret information, with all the weaknesses outlined above.

The best way to avoid these basic errors is to instruct your intelligence organisations to generate a non-secret understanding of your enemy, and the wider world context, from open-source information: from diplomats, newspapers, country experts, scholars, and conversations with normal civilians where you are operating. This understanding of the world should be privileged and given primacy for it is the context to which you must then later weave in your gleaming jewels of secret, and highly valuable, intelligence.

Most importantly of all, having a baseline common understanding of the world will help you ask the right questions of your intelligence agencies, whereas if you rely only on secret intelligence to build such understanding, you will ask erroneous questions that merely confirm your previous understanding (confirmation bias).

Once this information—secret and otherwise—is collated you as leader must create an environment where it is analysed in a way that minimises cognitive biases. The best way to do this is to allow competing voices in your intelligence organisations the space and freedom to make their arguments to you, and with each other. Allowing people and ideas to compete constructively is a tried and tested means of preventing biases from skewing your analysis. It can be improved further by drawing upon a diverse pool of intelligence analysts: deep experts in the region, and those with no expertise; language and culture specialists; military minds and civilians.

Put simply, competing diverse voices will ensure that the assumptions of one group of analysts will be tested robustly by another: this will improve your analysis. Furthermore, you will be less likely to frame the strategic problem incorrectly if you have multiple analysts coming up with different frames: your in-group out-group biases will be minimised.

Perhaps the easiest way to do this is to empower dissenting voices in your organisation. This can either be as a 'red team' whose sole job is to pick apart your intelligence analyses; or by allowing 'mavericks'—highly intelligent and creative oddballs—free reign to constructively criticise your analyses, or even come up with their own.

The failure to create and maintain diverse viewpoints among your intelligence analysis community can be very costly indeed. In the run up to the Iraq War in 2003, British intelligence was convinced that Saddam Hussein's Iraq had weapons of mass destruction but did not consider the alternative hypothesis: that he had already abandoned them some years before because of international pressure and UN weapons inspections.

This failure to create and empower competing voices was then compounded by the overconfidence of Prime Minister Tony Blair and his team, who argued that the intelligence was stronger than it was. The Chilcot Report severely criticised the British intelligence agencies for failing to correct this overconfidence: a classic example of obsequiousness bias.

Your main objective as leader is to establish that everyone in your intelligence structures listens to everyone else. Your patronage as leader is key to this—something as simple as turning to a dissenting voice in a meeting and asking for their opinion will empower them to speak up and send a message to everyone in your intelligence organisation that all voices are to be heard. Simple acts like this break down overconfidence and obsequiousness biases because your staff will be more confident in challenging the accepted wisdom.

Beyond these structures and actions, as a leader you must adopt a critical (as in reasoning, not negative) frame of mind as you consume intelligence. While you must keep an open mind as to what you are being told, the best starting point is slight scepticism. You, personally, must ask questions that test the assumptions of your intelligence organisations. Ask to see the original source report, if it is a critical piece of information. Ask to speak directly to the expert, or to the person who wrote the report, rather than the person who is briefing you. And if, after all this, everyone is still telling you the same story, then you must instruct them to reassess their assumptions. Competing teams rarely agree.

How to develop a realistic strategy

Once you have developed a sound strategic understanding of your enemy and the wider world context you can begin to form your strategy. This will comprise Ends (Objectives), Ways (Plans) and Means (Resources). Because these three elements are interdependent, and they all rest on a sound intelligence understanding, the formation of strategy is not a linear process, but a series of iterative conversations as you weigh each of the four elements up against each other.

Each conversation will have two types of people in it: those who set vision (usually political leaders), and those who have a deep understanding of the application of military force and the resources available to a particular country (usually generals or officials). You must be careful and deliberate throughout this process so that your top-level strategy remains in place for several years (the British strategy in the Second World War essentially remained unchanged for six years).

The first conversation you should have concerns your objectives (the Ends). You should consider these through the prism of your intelligence understanding of the enemy. If they seem realistic then evaluate them in light of the military and economic resources that your country has available (the Means). At this point you should only consider your resources in general strategic terms—do we have enough oil to fight this war?—rather than tactical: are the types of rifles we have of the right calibre?

Then weigh up whether you will be able to agree these objectives across your government. If you intend to fight with allies, this conversation must be happening not only within your government, but with your allies as well, until you agree. Finally, and significantly, you should consider the narrative of the war you are about to start. Will your public support it? Will the rest of the world accept or agree with what you intend to do? Will your friends and allies stand by you? Are you powerful enough not to care?

The answer to a later question (on whether you have allied support) will change your appreciation of an earlier question (whether your own public will back the war objectives that you are considering). During these conversations you will undoubtedly discover lacunae in your understanding of your enemy and the world, and so you must task your intelligence organisations to fill the gaps (and then revisit all the questions again).

As you debate, you will often find that the same people in your strategy team often adopt the same roles: one person will always come up with overconfident ideas; another will be more cautious; a third will have a keener insight into the enemy; a fourth will always focus on your own resources; and so on. This is natural as people tend to have stable personality types, and you should exploit this by creating a team where diverse viewpoints spark creativity. Your job as leader is to chair the discussion, keep the creative tension in the room, and make sure that your team keeps going around these conversations until all the questions have been settled.

If this process goes well, the resulting set of strategic objectives should be very high level (our aim is the total defeat of X country; our aim is to restore the pre-war borders of country Y); enduring (we can continue to pursue this aim for several years, if necessary); and supported by your own public, your allies, and enough of the world public at large. Conversely, a bad strategy will be low level and focussed on activity rather than outcomes (we will be launching an amphibious operation); short term (needs to be completed in weeks before our supplies run out); and unsupported (our own public won't like it, or our allies will shun us, or world opinion will be sufficiently negative that we may find ourselves sanctioned).

Once your objectives are relatively firm (the Ends), the next set of discussions are those that introduce the manner in which you intend to achieve your strategic objectives (the Ways), and to then reconsider whether you have the resources to carry out that specific plan (the Means).

At this point, you should start thinking creatively—ultimately, whatever your objectives, they will be achieved by influencing the mind of the opposing leader. For example, if your strategic goal is to evict the armed forces of your enemy from the territory of your ally, then rather than removing every enemy soldier from your ally's territory, could you launch an amphibious invasion of one of your enemy's cities to force them to withdraw their forces and sue for peace?

If you consider strategy in this way—that you are trying to force a change in your enemy's psychology—then your deployment of lethal violence becomes a way of signalling your intent, or of deceiving your enemy, or of sending messages to the enemy population. The resort to violence becomes part of the overall war of narratives, which is ultimately what war is about. Most wars are lost in the minds of leaders rather than on the battlefield.

As you consider your strategic plan and the resources you have at hand, you may find that you cannot achieve your objectives. Then you must either revise them, find further resources (or allies), consider a different plan, or a combination of all three. Finally, once you feel that you have a sound set of objectives, enough resources, and a realistic plan, you must systematically compare it to the intelligence picture that you have constructed—do your objectives and plan still seem realistic? Do you have enough resources? You must revisit these three factors—Ends, Ways and Means—until you arrive at objectives that are rooted in a realistic understanding of the world, that you can resource, and which are achievable.

There is no way of escaping this logic of forming a strategy and of balancing your Ends, Ways and Means. The most common error that leaders make—which is invariably fatal—is to have overly ambitious objectives when considered in light of the resources that they have to hand. You must avoid this pitfall at all costs.

Finally, you should consider too the differences between autocracies and democracies when it comes to strategy formation. You are unlikely to be able to change the type of government that you have, but you should be aware of their strengths and weaknesses.

Democratic systems are more likely to create organisations where people feel they can question assumptions and constructively criticise freely. That is, in theory, the standard human biases should be minimised by design in a democratic system. But this is not always the case: it is well recognised that in the Vietnam War, for instance, the US based its strategy on what it had learnt in the Korean War: supporting a 'democratic' government in the south, and not letting American troops go too far north for fear of antagonising China. But the assumptions underpinning these decisions were not truly questioned, resulting in a crushing strategic defeat for America.

Conversely, autocratic countries' strategies are more likely to be based on the whims of one person, usually a man. President Putin of Russia had clearly created a culture of 'yes men' surrounding him leading to his poor strategic judgement in invading Ukraine in 2022. Autocratic countries also tend to continue prosecuting failing strategies, because they are bound up with the leader who espoused them. But autocracies do have one advantage: they can set very long-term strategic goals, and follow them, providing the same leader stays in place. Democracies can and do change their strategies when an election ushers in a new administration: witness the Spanish government pulling its troops out of Iraq after the 2004 election.

So what does this look like in practice? In the Second World War, the Allies planned and carried out an amphibious operation on the Normandy coast in 1944. Although the overall strategic objective—to launch a seaborne landing to liberate occupied Europe—had been clear since the Europe First policy had been decided, practically every other detail was argued over.

At the highest level there were disagreements between the British, who wanted to prioritise operations in the Mediterranean, and the Americans, who pressed for an assault on occupied France much earlier (potentially in 1943). In the end, the discussion became a question of resources: because the Mediterranean theatre of operations already had Allied military deployments and operations ongoing, it was easier to reinforce their successes than to start an entirely new front in France (for which it would take time to build up logistics and personnel). The Allies decided first to assault Sicily, in 1943, followed by France the following year.

The two key people on the British side were Prime Minister Winston Churchill, and General Alan Brooke, Churchill's principle military advisor. Reading their recollections, they disagreed with each other vehemently about almost everything, yet each had great respect for what the other brought to the table (Alan Brooke: 'never have I admired and despised a man simultaneously to the same extent.'). One of the big things that they fought over was the focus on France, to the exception of launching other simultaneous attacks in Norway or Portugal. But they—and their staffs—also argued about other matters as well, such as whether it was better to use aerial bombing to attack German railways, or German oil supplies.

Each of these problems had to be worked through among large staffs of big personalities by considering whether each alternative plan contributed to the overall objective, and whether the resources were available for it. These deliberations had to be considered in light of the intelligence available—and further intelligence questions had to be asked and answered—such as whether the sand on different beaches in France would support the weight of different vehicles. The constructive debate of these issues over several months allowed the Allied commanders to tease out the assumptions underlying the different options and come up with a strategy that was ultimately successful.

The final element in D-Day planning was the element of deception. As soon as Normandy was settled on as the invasion site, the Allies set about convincing the Germans that the landings would be conducted elsewhere—in Norway, and particularly in Calais. An entire US Army Group was 'created' under General Patton, based across the English Channel from Calais, in Kent, complete with dummy vehicles, aircraft and fake radio traffic. Patton was chosen to exploit the enemy psychology: the Germans considered him the top commander on the Allied side—so wherever he was 'assigned' they would consider the main effort. Operation Fortitude was so successful that even several days after the Normandy landings the Germans were still diverting troops to the Pas-de-Calais.

Ultimately, the D-Day landings are the most complex, and successful, amphibious operations in history—and that is largely due to the quality of the strategic thought that went into preparing them.

A note on strategy

Strategy is not a nice art. Nor is it a place for wishfulness, thinking well of people, hope, or overconfidence. It is an arena where extreme realists excel. You should not be a strategic leader if you cannot contemplate sacrificing 4,000 of your own soldiers or allowing a civilian population to remain under extreme humanitarian pressure from your enemy—say, starving and under siege—to achieve a strategic objective. If you cannot make such tough decisions, you must allow others who can to lead. Lethal violence is ultimately a method of communicating in war, and if you are not comfortable with that fact, you should not be a war leader.

In 1940, a British infantry brigade—the 30th—was tasked with holding Calais, to divert German armoured divisions, thus enabling the evacuation of British and French troops at Dunkirk. The evacuation—eventually of 400,000 Allied soldiers—was fundamental in allowing Britain to continue to fight alone in the early years of the war. It saved a large part of her army. But enabling the evacuation required the sacrifice of the 4,000 troops of 30th infantry brigade. The night before they were overrun, killed or taken prisoner, Churchill telegrammed the Brigade commander, Brigadier Claude Nicholson, as follows:

> Every hour you continue to exist is of the greatest help to [our forces]. Government has therefore decided that you must continue to fight. Have greatest possible admiration for your splendid stand. Evacuation will not (repeat not) take place, and craft required for above purpose are to return to Dover.

You must examine your inner thoughts, in much the same way that individual soldiers consider whether they can kill or be killed, to determine whether you can make these least-worse choices better, and faster, than your enemy. It is not easy, and very few people have the right balance of intellectual strength, openness to different ideas, decisiveness, sense of personal responsibility, and ability to study detail as well as abstract their thinking to the highest levels. Reflect carefully about whether you do.

Often, so-called strategic leaders will see that their imperative is to save lives now, at any cost—via, for example, a peace treaty. Yet often they do so at risk of storing up further conflict, and hence further loss of life, in the future. The manner in which the First World War was inconclusively finished, thus sowing the seeds for the Second World War, is the best-known example of this.

Another way in which leaders may be tempted to save lives is through humanitarian action. When considered in isolation this is inherently a good thing. But as leader you must not confuse achieving humanitarian outcomes with your strategic objectives.

Humanitarianism can prolong wars—by keeping the families of fighters safe and fed, for instance—which ultimately leads to more destruction and loss of life. As leader, you must guard against being swayed by humanitarian imperatives in the media—sometimes saving lives now will entail much greater losses later.

Strategy is about settling political disputes, and if you find yourself fighting a war, then you have failed conclusively to resolve them through talking. As abhorrent as that sounds, you must remember that human society today, with its large, powerful, and mostly peaceful states is the product of millennia of wars past. This does nothing to reduce the pain and suffering that individuals, families and communities suffer in war—but it reminds us that war is an inherent part of human nature that we cannot wish away, as much as we would like to.

It is better that a war settles a strategic question, rather than ending with a peace that will once again erupt into war. War is a phenomenon that can shape outcomes that lead to a just and lasting peace, but if peace means the conflict is merely frozen, rather than settled, then it may be better to continue fighting. As a leader, your job is not always to avoid war at all costs: sometimes it will be the only way to settle a strategic question. And if you seek to avoid it, others will wage it against you. There is space between these two extremes, where a fragile peace can be made that can later be reinforced: ultimately the judgement as to which is which is yours alone as leader.

2.

LOGISTICS

Once you have a realistic strategy, the next most important factor in warfare is logistics.

Logistics in war means getting everything for the coming fight into your soldier's hands—a combatant who might be only a few metres away from the enemy. And what they will require encompasses everything that you could possibly think of—your army at war will need it all.

Most obviously, you must supply ammunition for weapons, and fuel for vehicles. You will need plentiful supplies of both for every hour of every day. Almost as important, but perhaps not quite as hour-to-hour, are food and water for your troops, and spare parts for all of your equipment: everything from oils and filters, spare tyres, batteries and hydraulic pistons, to generator parts, radio antennae and spare barrels for your weapons (many weapon barrels need replacing after a certain number of bullets have been fired through them).

Finally, beyond these critical supplies of fuel and ammunition, food/water and spare parts, are the more mundane, but still important items of supply. From trousers and underwear, to pens and paper, medical supplies, screwdrivers, tables, knives and forks, computers, nails, and pillows—if you can think of it, your military will have to supply it to your troops.

To give you an idea of how large most logistical supply systems are, the United States military supplies 7 million separate items to its troops, and the United Kingdom and France supply approximately 2.5 million items each. The logistics elements of most armies range from around 15 per cent to 25 per cent of total personnel, and a higher percentage of the vehicles. Successful militaries that expect to fight and win wars spend a lot of time and effort making sure that their logistics systems are efficient and effective, and able to survive attack by the enemy. Military logistics are mind-bogglingly vast, and extremely complicated.

In this chapter, you will read about military logistics from the factory to the front line, and why logistics are so important. You will then learn about the economic base required to produce weapons and vehicles, and whether you can substitute this for robust international alliances. The chapter then poses three questions: Are you able to transport these items to where you are fighting the war? Are you able to transport all these critical supplies along limited infrastructure the last hundred-or-so kilometres to your troops? And finally, can you protect your logistics system in a warzone if your enemy chooses to attack it?

The reciprocal of the last question is whether you can attack your enemy's logistics. This matters because it is much easier to blow up fuel trucks than to attack tanks. Furthermore, once you have successfully blown up the enemy's fuel tankers, immobile tanks, aircraft and ships will offer much easier targets. Each side is trying to do this to the other, and so as well as spending time finding and attacking your enemy's logistics, you will expend much effort in hiding and protecting your own.

Throughout the chapter, we'll assess some of the common mistakes that politicians and generals make concerning logistics, and scrutinise those wars and campaigns for which they created efficient and effective military supply systems. By the end of the chapter, you will understand the principles of military logistics and why logistical shortcomings are so constraining to operations. This will help you think about how to design your own military logistics systems so that they do what they need to in the wars you may embark upon.

* * *

In most depictions of war in literature or in the media, and even in the minds of many soldiers, logistics are considered boring. Indeed, logistics involve warehouses, spreadsheets, clipboards and paperwork. Much of the work of logistics involves checking what you have and don't have; counting things; and making sure that different items are stored and transported appropriately. In most militaries, logistics personnel are often looked down upon by their colleagues in, say, the infantry (in the British Army, they are unfairly derided as 'blanket stackers'!).

It is for a related reason that advanced democracies tend to have an inbuilt advantage over dictators when it comes to military logistics. This can be explained in one word: corruption. In a military logistics system, relatively poorly paid people—a private soldier might earn £20,000 ($23,000)—are responsible for looking after millions of items, totalling billions of dollars. In systems where corruption is the norm—and these tend to be autocracies rather than democracies—items of value disappear all the time, to be sold on the black market. Higher value military-grade equipment is replaced with cheap quasi-equivalents. And at the extreme, entire ghost units are created, and the value of the war materiel that would be supplying them is stolen.

If you have a corrupt government, yet you wish to be militarily successful, you must consider how corruption erodes your ability to supply war materiel effectively to your armed forces.

Dictators also tend to place a premium on what we might call vanity weapons systems: big missile launchers, the latest technology, and specialist capabilities like ships and satellites, rather than 'boring' logistics. In short, autocrats like the symbols of military power, because they are by definition 'strong men' who like big weapons. It is true that sometimes democratic leaders fall for these ego-boosts, but they are more constrained by parliaments and public opinion. Yet if your country is serious about fighting and winning wars, you must spend a great deal of time focussing on container ships, railways and articulated lorries, and systems for counting everything.

Boring or not, logistics are fundamental to military success. Quite simply, this is because without fuel your tanks are merely expensive stationary targets; without ammunition your artillery pieces are a burden; and without food and water, your soldiers will rapidly degrade and become ineffective in combat, making them more likely to be captured or killed. In war, getting supplies of ammunition, fuel, and spare parts to the front line is far more important than the number of tanks that you have at that front line.

It is because of this logic that all successful military operations are planned around logistical constraints, and these can be summed up in a simple question: At each stage of your operation, are you able to supply your troops with what they need? If, at any stage of your operation, the answer is 'no', then you must redesign your operation. President Eisenhower, previously the Supreme Allied Commander of the logistically hugely complex 1944 D-Day operations, articulated best this process of looking at military objectives in the light of logistical planning constraints when he said, 'Plans are worthless, but planning is everything'.

The D-Day landings were undoubtedly the most impressive military logistics operation ever launched. On day one, the Allies landed 156,000 troops to secure the beaches of Normandy.

As soon as this was done, two prefabricated Mulberry Harbours were towed across the Channel to land supplies in the interval before major French ports could be captured. All told, the Allies unloaded 2.5 million men, 500,000 vehicles and 4 million tonnes of supplies through one of the Mulberries over its ten months of operational use (the other was destroyed in a storm shortly after deployment).

In another piece of extraordinary innovation for the time, underwater pipelines were developed to deliver petrol, diesel oil, and other lubricants to the Allied expeditionary force on the continent (liquid hydrocarbons took up 60 per cent of the supply needs by weight of the Allied expeditionary force). In total, seventeen pipelines were developed in an operation known as Operation PLUTO or PipeLine Under The Ocean, and they accounted for nearly 10 per cent of the supply to Allied troops.

Before going into the nuts and bolts of a military logistics system, let's have a brief look at the scale of logistics required to fight a war on land.

The scale of logistics in war

The logistics tail of a modern army is very long indeed and makes an easy target—the more technologically advanced your force, the greater the ratio between combat units and support units (including not only logistics but also medical, intelligence, engineering, mechanics, technicians, etc.). This ratio is known as the Tooth-to-Tail Ratio and in a modern-day armoured campaign can reach up to 1:10 or 1:15—that is, for every person whose primary job it is to fight the enemy, you will need ten or fifteen whose job it is to support them. In the First World War this ratio was just 1:2.5, and in the Second World War it was only 1:4.

The four most important things that you must transport to the front line are ammunition, fuel, spare parts, and food/water.

Exactly how much of each depends on two main factors: the intensity of combat—how many patrols, engagements, battles, and assaults you intend to conduct—and the types of military unit in your army, as each one has different operational requirements in respect of these four main items.

The volumes of these four essential supplies required has changed throughout the twentieth century. In the Second World War, the US armed forces used about 1 gallon (3.8 litres) of fuel per soldier per day. With greater mechanisation and use of air power, this had increased to 4 gallons per soldier per day by the first Gulf War in 1990. A mere twenty years later, in the Iraq and Afghan Wars, fuel consumption reached 16 gallons per soldier per day (some 71 per cent of which went to the air force).

Similarly, because of general increases in firepower over the twentieth century, ammunition as a percentage of total supply has gone from 12 per cent in the Second World War, to around 25 per cent by the late 1980s. Conversely, the percentages of total supply allocated to food and spare parts has steadily decreased (although the Russo-Ukraine War of 2022 onwards, with the vast scale of Russian vehicle losses and damage, may temporarily reverse this trend).

Beyond these figures lies a simple truth. If the war on which you are embarking lasts longer than a few months—and mostly they do—you must consider a form of total national mobilisation of the economy and population to harness them to production of war materiel. Factories must be nationalised, infrastructure controlled, personnel directed into jobs and tasks. Whether you are able to do this successfully depends upon the degree to which your population supports you in your war aims, and so sometimes this is easier for democracies than autocracies.

For example, Finland completely mobilised its nation during the Second World War when fighting the Soviet Union, enabling it to resist and fight the Soviets long enough to survive and remain an independent country. It is notable that whilst not put into practice currently, many of the national mobilisation laws remain on the statute book.

It is beyond the scale of this book to describe this type of national mobilisation and resilience. Suffice it to say, if you get involved in a long-lasting war the degree to which you are able to harness your entire nation behind the project will be a significant factor in your victory or defeat.

So how much will your soldiers require in the field?

At soldiering's most basic, light role infantry—what you might imagine as 'traditional' soldiers equipped with rifles and machine-guns, and moving on foot—require the least logistical support. At the other end of the scale, your armoured units—including tanks, armoured personnel carriers and artillery—require truly massive levels of supply. Artillery especially, when one considers the weight of each shell, and the rate of fire (number of shells per minute), will dominate the logistics requirements of your army. It is for these logistical reasons that so few armies can successfully field expeditionary armies comprising armoured and artillery forces.

Your average light role infantry solider—not a specialist and trained to operate as part of a standard infantry company on foot—requires supplies of ammunition and food and water. Food is the easiest to estimate—a British Army 24-hour ration pack weights 1.8kg. Estimating water requirements is harder—it depends on the climate and whether there are local water sources. Let us say that, on average, each soldier needs 5 litres (5kg) of water a day for drinking, washing and cooking.

Ammunition requirements are difficult to estimate because usage depends entirely on how many combat hours your soldiers are engaged for. On some days they may not fire a single bullet;

on others, they may be in contact with the enemy for twelve hours and fire thousands of rounds. So let us assume that on a typical day your average soldier fires 300 bullets, which is ten magazines of ammunition. Each of them could get through that amount in a medium-scale firefight lasting less than an hour.

Bullets vary in size and weight but the NATO standard 5.56mm rifle bullet weighs approximately 15g. So our average infantryman's daily usage of 300 bullets weighs 4.5kg. The ex-Warsaw Pact armies, or forces that rely on AK-47 style rifles, use the 7.62mm (weighing 25g) bullet which means that their soldiers can carry fewer of them for the same weight. NATO deliberately adopted the smaller/lighter 5.56mm round simply because it meant that each soldier could carry more bullets.

To these calculations, you must factor in more ammunition weight for the heavier weapons that the infantry company might carry—perhaps lightweight mortars, or heavier machine guns (even light role infantry companies will carry a selection of these heavier weapons). Two hundred rounds of 7.62mm machine gun ammunition weighs about 5.5kg (including the clips that link each bullet to the next thus creating a 'belt' of ammunition to feed through the machine gun). This is what each of your soldiers would carry on a patrol to contribute to the heavy machine gun carried by one of them. Each (51mm) mortar bomb weighs about 1kg—let's say your soldiers will carry one mortar bomb each on average to contribute to their comrade's mortar tube.

With food, water and ammunition, you have already calculated up to 17.8kg per person per day. To this we can add other essential items: batteries, food, fuel, replacements for clothes, camouflage cream, soap and razors. On average these add up to 700g of personal items per day.

These rough calculations bring you to a total of 18.5kg per soldier, per day. In high intensity combat, fighting all day, this would be a serious underestimate, but it is a useful figure when taken as an average over a long campaign. Supplying only the soldiers in a battalion of 500 troops (excluding vehicles or other equipment such as generators) involves 9.25 tonnes of materiel per day. Scaled up to a fictional light role division of 10,000 troops, you will need 185 tonnes. A 40ft shipping container can carry 26 tonnes, so your divisional requirements are seven shipping containers per day just to supply your soldiers with the basics they require to survive and fight (in reality the supply demands would be much bigger with equipment and vehicles).

At the other end of the scale is one of your armoured divisions (of 16,000 troops). The biggest difference is the scale of the fuel and ammunition requirements. For example, during the First Gulf War in 1990, a typical US armoured division was made up of 350 tanks and 200 armoured infantry fighting vehicles. These formations required 5,000 tonnes of ammunition, 550,000 gallons of fuel (1,485 tonnes), 300,000 gallons of water (1,140 tonnes), and 45 tonnes of food, amounting to 7,670 tonnes *per day* (295 shipping containers). Nor does this include the substantial spare parts required.

To give you some idea of the overall scale of logistics in the Gulf War: British, French and US forces deployed five armoured divisions between them, and six other non-armoured or light role divisions. By the figures laid out in the preceding paragraphs, this is over 1,500 shipping containers of supplies *per day.*

Finally, artillery. It is extremely hard to judge the logistics requirements of your artillery force because everything depends upon rates of fire—how many shells are fired per minute, from how many guns—and the size of the shell, rocket or missile (missiles are simply rockets with guidance systems). We will look at some of the different types of artillery in Chapter 5, but for an overview of your artillery logistics requirements consider the 155mm howitzer, the mainstay of many armies. One 155mm shell (and the explosive charge to fire it from the gun barrel) weighs 50kg.

A sustained rate of fire during an artillery barrage might be two shells per minute making a total of 120 shells per hour. This runs to 6 tonnes of ammunition per hour, per gun. Normally, there are eight guns in a battery, so 48 tonnes of ammunition would be required per hour, not including fuel, food/water, or spare parts. Finally, the 155mm system is one of the lighter artillery systems available to most militaries; even so, it requires approximately two shipping containers of ammunition per hour to operate a battery of guns at a normal rate of fire.

These numbers mean that if you are serious about having armies that will fight and win wars, each artillery regiment must have its own paired logistics regiment that serves only that unit. For your artillery forces to deliver effective firepower on a modern battlefield, approximately 50 per cent of their vehicles and personnel need to be focussed on logistics—getting the ammunition to the gun line through contested territory. Professional armies that are prepared to fight and win wars will have this ratio or something close to it—anything less will mean that either the artillery can only operate in fixed positions next to (for example) railheads, or it simply cannot deliver artillery firepower in combat.

The daunting statistics of military supply for different fighting units mean that great effort is expended reducing the supplies required by armies in the field, because it can make the difference between defeat and victory. One of the most innovative approaches to this objective was taken by General Bill Slim, Commander of the 14th Army in Burma (Myanmar) in the Second World War. Working with his logistics staff, the supply requirements for the 14th Army Division were reduced from 400 tonnes per day to only 120 tonnes. Vehicles were ditched in favour of pack animals; tents were made from silk rather than canvas; food rations substituted for lighter versions.

These logistical constraints cannot be wished away—they are part of the arithmetic of warfare. One of the key mistakes that political leaders make repeatedly is to think that the conflict upon which they are embarking will be shorter than it eventually turns out to be. This means that they only have stocks for, say, a two-week conflict, when it turns out to last months, or years. Then, new stocks need to be manufactured or procured from your allies—usually they cannot be bought off-the-shelf. This supply crunch then affects your on-the-ground commanders who will have to make sub-optimal decisions because they know that they are about to run out of fuel, or that they will have to severely ration ammunition. Ultimately the logistical constraints start to dictate the military operation.

This trend will become even more acute when it comes to very expensive or rare stocks. A common example of this in modern warfare is supplies of precision guided munitions (PGMs)—also known as laser guided bombs—that you can drop precisely onto targets from aircraft thus minimising ammunition usage and collateral damage. PGMs are expensive—a single Paveway laser-guided bomb costs £22,000—and so they are rare. In the recent war in Ukraine, Russia became unable to replace PGMs very early on, and so lost its ability to hit moving targets like Ukrainian logistics trains; the United Kingdom—a much richer nation than Russia—ran out of Paveways during the conflict in Libya, and had to ask the Americans for an urgent delivery.

These logistical constants to warfare are a useful way for you to assess whether your army will be successful in what you are setting out to achieve. Is your force formed of approximately 25 per cent logistics units? Is your artillery supported 1:1 with logistics? Are your stocks of ammunition and fuel sufficient for a long war? Do you have supplies of critical components like microchips?

There are further questions: Does your country have a manufacturing base for weapons and supplies? Can you get your war materiel to the warzone? Is the road and rail network in the warzone sufficient to get your supplies to where they are needed? And finally, are there enough combat units to protect your logistics? It is to these later questions which we now turn.

Producing war materiel and getting it to the warzone

To supply a piece of equipment to your combat soldier you must have access to the right raw materials. The materials required have changed throughout history as different technologies of war have dominated the battlefield. For much of the post-industrial revolution era, coal, iron (steel), and later oil have remained critical supplies. Once vehicles became common, rubber for tyres became especially important—in the Second World War, Japan managed to capture most of the world's rubber production in South-East Asia, pushing the United States to enforce a rubber-saving speed limit on its roads, while synthetic rubber was invented to reduce dependence on the real thing.

In the twenty-first century, modern military equipment is much more reliant upon information technology. In addition to explosives, weapons, and chemicals, your military can be brought to a grinding halt by cutting off its supply of microchips, lasers, software, and specialist communications or navigational devices.

As the century progresses, and information technology is further integrated into military equipment, other resources will become increasingly important in its production. Chief among these is the 'rare earths', a group of seventeen heavy metals that are used in electronics, but particularly in batteries and (computer) screens. It is worth noting that China has 38 per cent of rare earth deposits globally, Russia has just over 10 per cent, and the United States merely 1 per cent.

It is because of this military reliance on certain items that many countries restrict their export as a matter of course. For example, the British government has a list of items stretching over 300 pages—from explosives to nuclear isotopes, and from bullets to gyroscopes—that it bans for export unless specific permission is sought. It is also for this reason that economic and/or trade sanctions are often used as a tool of foreign policy to damage the economy of the sanctioned nation, or to stop them accessing the materials or items that they require.

Sanctions are often controversial—economic sanctions hurt the population as much as, if not more than, the government—and their effectiveness can be limited, because there are other countries that will refuse to sanction their allies and continue supplying proscribed items, thus creating 'back doors'.

Sanctions do work well in the form of arms embargoes that stop militaries or governments from accessing high-end or heavier military equipment. A good example of this is the arms embargo on Somalia, in place since 1992, which has meant that the conflict, although continuing since then, has remained contained within Somalia's borders, and is being prosecuted with small arms, rather than with tanks or artillery. In the (ongoing at time of writing) war in Ukraine, there is sound evidence that restricting Russia's access to microchips and other advanced avionics has limited its ability to produce its own precision guided munitions, contributing to a supply shortage.

Assuming that you can secure a supply of the commodities or components needed to produce military items, you will also have to ensure that you can produce enough of your own weapons and supplies, at the rate that they are used in a conflict. If you cannot produce such military equipment yourself, you must source it from close allies that you are certain will continue to supply you, even in an adverse international environment where other powerful countries decide to impose sanctions.

Hence redoubtable military powers always try to keep production of specific military items, especially high-tech ones like submarines and missiles, onshore because this not only maintains the pool of expertise required within your own population, but also avoids having to rely on others to supply your weapons. And it is also for these reasons that factories that produce key bottleneck items are often the target of strategic air raids—in 1943, the Allies bombed Nazi Germany's ball-bearing factories, because they had worked out that ball-bearings were an essential item in a wide variety of military equipment such as planes, guns, tanks and submarines. Can you produce critical weapons systems onshore, and are your supplies of crunch components protected?

Once you have secured the raw materials, and produced or sourced your weapons from allies, you will need a large nationwide defence logistics organisation that ensures adequate stocks of weapons, ammunition and war materiel are maintained. You must maintain a delicate balance between keeping the costs down during normal peacetime activities like training and ensuring that there is enough equipment and supplies, especially ammunition, for emergency deployments of military force. As we saw with Russia in Ukraine and the United Kingdom in Libya, this is not always a balance that is achieved.

The national defence logistics organisation in your country is the interface between industry—that creates the equipment, ammunition, fuel or other militarily required items—and your military. It will take the military's wish list, and then tell industry how much and of what to produce. It will then prepare these items and organise them for onward transport to your military in the field.

Key to this operation is a system of categorisation like the NATO Stock Numbers (NSNs). These are 13-digit codes classifying every item into a category and a country of origin within the NATO alliance. This enables an infantry soldier in a trench to tell his lance corporal that he needs 1x NSN 8420–01–112–

2889 (thigh-length male boxer shorts, size 50, made of cotton, from the US) because he has split his previous pair. That number is standardised across the entire system, eventually meaning that your factory will manufacture another pair of boxers to replace the one given to your soldier from the stocks held at the front line.

Finally, within defence logistics, you should make great efforts to standardise vehicles and equipment so that generic parts/supplies can be used across different types of equipment. Most obviously, you should seek to standardise ammunition—as we have already seen with the 5.56 rifle bullet and the 155mm howitzer shell—and fuel (which represents the largest single item of supply for a modern military).

This idea of a single fuel was championed by the United States military in the late 1980s. The Americans developed JP-8: a kerosine-based fuel with added anti-icing, anti-corrosion, extra lubricants, and anti-statics (to make them less flammable when stored)—everything needed for a fuel to be used in different climates, stored for long periods, moved around a lot and that might come under enemy attack.

Although using one fuel for many different engine types may hamper performance, and can make some engines wear faster, these difficulties are vastly outweighed by the logistical benefits: the single fuel can be used in everything from aircraft to tanks, and Humvees to generators. It can heat buildings and power cooking stoves and even be used as a coolant in some engines! This idea has now been mostly adopted across NATO; by comparison, the Russian armed forces still use discrete fuels for different types of vehicles, putting them at a severe logistical disadvantage. If you are using different fuels for different vehicles, you should consider standardising.

Once you have produced your weapons, equipment and supplies, and organised and categorised them ready for your forces, you must transport millions of tonnes of supplies to the warzone. Inevitably this means transport by sea, or a combination of sea and road/rail.

A fascinating recent example of this is the war in Afghanistan from 2001 to 2021 where up to 150,000 troops from a coalition of developed countries fought the Taliban insurgency (and lost). Afghanistan is a landlocked country, so supplies either had to be shipped and then unloaded and transported across land, or flown in by air. The air route was ruinously expensive—ten times the cost of the sea and land option—and so was limited to supplying ammunition (the mostly night-time flights were known as 'bang runs'). The military logistics required were immense—3 million litres of fuel per day.

Supply was predominantly by two routes. The first was from the port of Karachi, across Pakistan, and through two mountainous passes into Afghanistan along approximately 1,700km of road. At times, up to 400 trucks per day were passing through Pakistan, but this handed a huge amount of power to the Pakistani government who were able to cut off or limit the Coalition's military supplies. Hence in 2009 another route was established from the port of Riga in Latvia, across Russia, Kazakhstan and Uzbekistan across 5,169km of railways. Up to thirty trains a week began to move supplies into Afghanistan, eventually accounting for one-third of the total supply required.

Strategically speaking, the ability to supply your military when it is deployed relies on control of, or safe passage through, the world's oceans—you cannot fight a land war if the country that controls the oceans doesn't want you to (unless the country you are attacking is next door, as with the Iraqi invasion of Kuwait in 1990). It is no coincidence that all the world powers since the 1500s (the Portuguese, the Spanish, the Dutch, the British, and the Americans) have been maritime powers with global navies able to control the sea. You must keep your shipping lanes open to protect the flow of supplies to the theatre of war. If you cannot do this, you will lose the war.

If you can guarantee safe passage through the seas, you must seek to have sufficient shipping available to transport vast amounts of supplies. In the 1982 Falklands War, the British found themselves severely short of vessels, and had to requisition fifty-four ships from the British civilian fleet to transport supplies the 7,000 miles between Britain and the South Atlantic. This included fifteen oil tankers, and a P&O cruise liner was even converted into a hospital ship for the wounded. In the event, these requisitioned ships carried 100,000 tonnes of supplies, ninety-five aircraft, 9,000 personnel and 400,000 tonnes of fuel—there is simply no way the invasion could have occurred without them.

The largest scale example of a 'shipping lane' conflict is the Battle of the Atlantic in the Second World War. Also known as the Tonnage War, it was a naval theatre of operations lasting almost throughout the conflict. It consisted of an Allied naval blockade of Germany to halt the supplies that it needed for its war effort, and a counter-blockade by Germany that was largely aimed at interdicting military supplies coming from the United States to Britain and to Russia.

Throughout the war, but particularly in 1939–42, these supplies were essential in keeping the British and Russians in the war. Britain, for example, required a million tonnes of supplies a week to survive and continue fighting. Without defeating the German submarine (U-boat) threat to Allied maritime supply, there was no possibility of altering the conflict from a largely defensive war, to an offensive one where Nazi Germany could begin to be pushed back.

As we know, the Allies won the Battle of the Atlantic, and the Second World War. But it came at some cost, accounting for 72,000 men, 3,500 merchant ships, 175 warships, and 741 aircraft before they were able to subdue the threat, eventually sinking 783 German U-boats. This was a logistics war at the strategic level which had to be won before the Allies could prevail in the wider conflict. As Winston Churchill said at the time: 'the only thing that ever really frightened me during the war was the U-boat peril'.

We now turn to logistics at the tactical level: once you have acquired your resources, built your weapons and equipment, and delivered them to the theatre of war, you then have to transport your supplies through the warzone, all the way to your soldier standing a few hundred metres from the enemy.

Logistics inside the warzone

When military logistics enter a warzone, they become very fragile. This is not only because you are trying to move large amounts of supplies over infrastructure—roads and rail—that is simply not designed to take the heavy loads required, but also because your adversary will be blowing the bridges, mining the road junctions and sabotaging the railway yards that you rely on. In addition, during the fighting the enemy will be attacking your supplies before they reach your troops—with the priority of destroying your fuel first, and then your ammunition. You will be doing the same to them.

There are a number of basic principles that apply to logistics within the combat zone.

First, you must hold your supplies as far forward as possible while still being safe. Holding your supplies forward also means that they are likely to be positioned near where the vast bulk of your combat units are located, hence you will have assets on hand to help protect your logistics vehicles and supply. Logistics vehicles are only very lightly armoured, if at all and, at most, carry a small calibre machine gun for protection against light role enemy infantry only.

Second, you must try to predict what your requirements will be in the days and weeks ahead as the intensity of combat waxes and wanes. This is achieved by constant communication between the fighting and logistics parts of an army about what their stocks and requirements are. This communication is immediate: for instance, once an infantry squad has taken an enemy position, the first thing they do is arrange their own protection—that is, which soldier is covering in which direction. The second is to calculate their ammunition usage, and how much more they need in the immediate period, be it 24 or 48 hours. They do this even before moving casualties.

Third, and related to your ability to assess your requirements accurately, you must attempt to keep your logistics moving, because mobile targets are much harder to attack (this is also why precision guided munitions are important). As soon as your logistics tail grinds to a halt, it becomes a target, and needs protecting. Camouflage it, or otherwise deceive the enemy about its presence before it soaks up further combat power for protection, thus further reducing your ability to advance.

This principle means that in a fast-moving war, your combat units and logistical tail resemble a battlefield concertina as attacking forces advance, and then pause to allow the logistics to catch up and conduct resupply. Coordination of the two elements is of pivotal importance because running ahead of your supply is a sure-fire way to lose both.

You also must avoid the opposite problem: traffic congestion near the front line and tailbacks tens of kilometres long are not uncommon and make an obvious target for enemy forces. In wartime one of the roles of the military police is to direct traffic, ensuring that it remains free flowing and avoids becoming concentrated in one area, clogging it up. This is also why—as we will explore in more detail in Chapter 9—retreats are so difficult to execute: an army's logistic support must be thinned out so that fighting units can retreat through the area, but it is hard for them to fight without logistic support.

Fourth, and finally, you should think of supply not only in terms of supplies and equipment heading towards the front line, but also in the opposite direction. While fuel, ammunition and other supplies need to proceed ahead, casualties (whether dead or injured), prisoners of war, and damaged and faulty equipment that can be fixed must all be moved to the rear. You should also consider mortuary units to pick up the remains of your soldiers, to return for identification and burial.

This reverse flow is vital to the medium-term functioning of an army. In the case of casualties, all troops will benefit from better morale if they know that there is medical support that they can access, or that their body will be cared for. Prisoners are a valuable source of intelligence; treating them well will help justify your war's narrative to the wider world. Broken vehicles, once fixed, will re-supply your armoury, and once removed from the front lines will not block roads or take up precious space on your rail networks.

How do these principles fare when applied to the battlefield?

Your army will be tied to the logistics tail that runs from a sea (or river) port, along a railway and/or a road. These are known as Main Supply Routes or MSRs. Even if your vehicles can operate cross-country—this is why tanks have tracks rather than wheels—98 per cent of your logistics vehicles will be tied to roads. So if fighting occurs away from the MSRs, then your units must return again and again to that network to get re-supplied. Because rail networks do not go everywhere, the last stretch of military supply—and often the most dangerous—is always along roads.

This explains why so much fighting occurs along roads, and why military engagement in dense forest or across mountains is so taxing—not because your fighting units cannot get through, but rather because it is extremely difficult to supply them once they do. Some armies—like the Russian Armed Forces for instance—rely heavily on railways for their military logistics, but while excellent in a defensive battle, they have obvious limitations if you are trying to take territory, as this entails laying your own rail network.

Given the importance of the road network, your military will have a system for classifying roads (and bridges, junctions, etc.) and vehicles into weight categories and limits. Marked on a map, this will enable you to see whether you can manage to transport the 295 shipping containers per day required for your armoured division to keep fighting. It will certainly dictate where you can position your artillery batteries.

Your classification system will categorise roads and vehicles by the loads they can bear without degrading. Most highways in Europe can carry a maximum of 44 tonnes (in an articulated lorry with the load spread across several axles). This will enable you to transport a shipping container of supplies. But a main battle tank can weigh anything from 46 tonnes (the Russian T-90) to 64 tonnes (the British Challenger 2), which would soon degrade the road network, necessitating military engineers to repair or improve it.

Once you have classified the roads, you must consider bridges and road junctions which are normally far more constricted in the weight that they can carry. For example, the Golden Gate Bridge in San Francisco is limited to 36 tonnes and would be damaged if you drove your tanks across it. Your engineers might need to build another bridge if you wanted to invade San Francisco from the North. Other famous bridges are even more restricted—Tower Bridge in London can only take 18 tonnes—and you would be unable to get a shipping container across without breaking it down into smaller loads. Junctions, which come under greater stress when vehicles turn on them, often present similar pinch points.

These realities mean that you must designate every road in the warzone with a load classification. This is also true of the vehicles in your fleet, which will usually be marked with their own weight, and in the case of logistics vehicles, the size of loads that they can carry. If you ignore these limits, your vehicles will suffer maintenance problems, and the road network will start to fall apart, jeopardising resupply of your combat forces. You will find that, frustratingly, roads may have an acceptable load capacity, but there will be a bridge or a junction which acts as a bottleneck. One of the key jobs of military engineers is to survey any bridges or junctions early on to see if they need to be reinforced or replaced.

You will have at your disposal various military logistics vehicles, with the lighter, smaller ones being able to go further forward. You will also need heavy equipment transporters to move your tanks and other tracked or heavy vehicles long distances. These 'tank transporters' might be up to 80 tonnes in weight. Because of their size, you should only bring these vehicles to the edge of the battlespace.

At the next level down, you will rely on two workhorse vehicles: a lorry to carry shipping containers, and a fuel/water tanker. These large logistics vehicles travel across what are known as 'rear areas' of the battlespace, and so, although not expecting to come face-to-face with an enemy unit, you will still need to allocate combat units to maintain security in these areas—primarily to ensure that your logistics are not disrupted.

Approximately three soldiers per kilometre of road are required to protect a supply line from moderate partisan or insurgent activity, with many more required to defend key nodes like bridges and junctions (the three soldiers are not spread out over the kilometre but grouped into platoons and companies that patrol the route). Over a large battlespace this can add up to thousands of troops, cutting back the numbers that you have available for combat. Moreover the further forward you advance, the longer the supply lines that you must protect—this is simple mathematics that many overlook. You should consider it carefully.

The first of your workhouse vehicles is a lorry that can transport a shipping container, in total weighing about 40 tonnes. Many armies have designed vehicles that can pick up and drop off shipping containers by themselves with a mini-inbuilt crane. As a result, battlefields are often littered with disused shipping containers, which are often repurposed as convenient temporary shelters or command posts.

The second workhorse of your land army is the fuel tanker. These can generally carry around 20,000 litres of fuel, or 30,000 litres of water (the latter means a total vehicle weight of approximately 35 tonnes). To put that in the context of fighting power, each of your armoured divisions will require 105 fuel tankers and thirty-eight water tankers per day to keep moving and fighting. Every army suffers from these constraints: the US Army had to briefly pause its advance to Baghdad in the Iraq War of 2003 because they could not bring up fuel fast enough.

Your two workhorses will deliver supplies to divisional-level (10,000 people), and probably down to brigade-level (3,000 people), resupply areas. Here you should make a distinction between mechanised and armoured fighting units, or other units who can transport themselves to the resupply areas, and other units—usually infantry—who are unable to move because they are holding ground or not very mobile.

Mechanised and armoured units will travel to the resupply area themselves. These areas—and the two workhorse logistics vehicles—will usually be kept out of the range of direct fire from rifles and machine guns, yet close enough to the front to conduct a resupply every 24 hours. This gives us a minimum distance between front and resupply area of around a mile, and a maximum distance of 10 or 12 miles if your armoured columns are moving at pace.

As far as possible, you will need to keep moving the brigade resupply areas, otherwise they will be targeted by the enemy. If you expect to be there for more than six to twelve hours, then think about camouflage, deception and protection—particularly from air and artillery attack, as one direct hit could set off all the ammunition and fuel (the majority of your supplies). In a fast-moving battle, your air defence assets will be in very short supply.

After the last mile, the logistics units of your army will hand over supplies to the fighting elements holding fixed positions—usually infantry. Fighting units are unable to come to resupply areas, and so their own integral logistics teams within the battlegroup will collect the supplies and then distribute them to their own fighting companies, after breaking them down into smaller loads that fit on 7–14-tonne lorries (a battlegroup of 1,000 soldiers require two lorries of this size per day, for food, water and ammunition). The fighting unit must provide the protection for its own logistics moving across the battlefield.

Your battlegroup should aim to hold 24–48 hours' worth of supplies, as well as technical stores like large generators and spare parts for critical equipment. These supplies would be delivered to each of your fighting companies, which again has its own logistics element, usually run by a senior non-commissioned officer.

At company level—of 150 troops—the aim is to conduct replenishment during the battle—this predominantly means supplying ammunition and moving casualties rearwards (if there are no casualty evacuation helicopters available). Your army will need a four-wheel drive car, tricycle or motorcycle with trailer to achieve this. As an example, the British Army uses all-terrain quad bikes with trailers that deliver ammunition forwards and take a casualty stretcher rearwards. Casualties then travel back through all the levels of the logistics system, usually carried in a battlefield ambulance, to a field hospital.

This is the most intimate level of logistics support available—beyond that, your troops will be carrying their own ammunition, food and water into battle. And in combat, the higher the morale of your troops, the more likely they are to survive the chaos, bloodshed and destruction of war. The morale of your troops is the subject of the next chapter.

3.

MORALE

Once you have built a realistic strategy and planned your military logistics from raw material to bayonet, the next most important thing is to prepare your troops such that they have the morale to fight and win. Morale is much more than simply happiness. It is the glue that keeps your force together.

To put it starkly, close combat is among the most frightening of human experiences. Each nearby explosion, or crack of a bullet, can strike fear into even highly trained soldiers. The death of a comrade, or a close friend, can have a shattering impact. Not sleeping for two or three days, or carrying injuries, will wear down the strongest of people, eventually breaking down teams as individuals become withdrawn and focus on their own survival.

Good morale is the answer to these unavoidable problems: it inoculates your troops against fear, and your teams against disintegrating. It is such an important ingredient in warfare that a well-equipped force with poor morale will almost always lose to poorly-equipped troops with high morale. And ultimately, in close combat, only those with the utmost will to survive and win will prevail.

General Bill Slim in the Burma Campaign of the Second World War understood this keenly. He was a real soldiers' soldier, who toured all over the battlefield visiting units and talking to the troops whilst constantly repeating the maxim that the Japanese forces could and would be defeated. He reinforced this with firm-but-fair discipline so that soldiers knew where they stood. The resulting high morale sustained the 14th Army until victory over the Japanese forces in 1945.

Morale is so important in generating and maintaining your ability to fight that your enemy will explicitly seek to destroy your morale, thus shattering your cohesion and eroding your fighting spirit. Indeed several armies make it a central part of their doctrine, and you should too. The British Army's doctrine, for instance, states that its main aim on the battlefield is to shatter the enemy's cohesion and will to fight. More battles are won and lost in the mind than on the field of combat and if you can get your enemy to surrender or run away, that is much easier than killing every last one of them.

So, cohesion—keeping your small teams together and functioning well—is critical in close combat where battles can be decided by a small number of troops holding vital terrain like a hilltop or a bridge. Thus the morale of your force, and the cohesion that this creates, becomes the lynchpin of military success.

There are other more subtle reasons why good morale is important. It gives your troops a feeling of much greater energy and optimism which they will draw upon when events inevitably go against them on the battlefield. Troops with good morale feel like they can achieve anything, and overcome any odds. They are more likely to work for their comrades rather than themselves, are less prone to become injured, or if injured they will cope better.

And perhaps second only to the team cohesion that it maintains, troops with high morale are more likely to be disciplined in carrying out the boring repetitive tasks of soldiering. Do they sweep the ground by their vehicle for mines before stepping down? Do they pause their patrol regularly to stop and listen? Do they clean their rifles every day? Being disciplined about these matters is what keeps soldiers alive because they won't step on a mine unnecessarily, and nor will their rifles jam when they need them most. Put simply, there is an unbreakable link between morale, discipline, and casualty rate.

I witnessed this myself first-hand. When soldiering in Afghanistan, I had to travel across the battlefield to different units. This gave me a keen insight into what made them professional and effective, or otherwise; that is, good and bad units. Without fail, bad units had weak leaders, whom their troops did not respect, and who spoke about them behind their backs. Nothing was more corrosive to morale. This lack of pride in themselves and their leaders meant that they often couldn't be bothered to do things that were irksome or difficult. In turn it meant that vehicles would take the same routes at the same times of day; patrols would not bother to put protection out on high ground when moving through obvious ambush sites; and checks for roadside bombs—the responsibility of everyone from private soldier to general—were cursory and missed the telltale disturbed ground, or simply neglected entirely. Sadly, these units took higher casualties than they might have otherwise.

Morale is vital: let's turn now to how to build and maintain it in your forces.

The high-level foundations of good morale

The first consideration when trying to create a fighting force with high morale is quasi-philosophical. It requires you to think deeply about how your country is structured. This is because the foundations of good morale are nested in a relationship of mutual trust and respect between your military, your government, and your people. Each needs to be reflective of, and responsive to, each of the other two elements and there should exist a positive tension between all three. What does that mean in practice?

First, a government must have the support—whether democratically elected or otherwise—of its people in prosecuting strategic ambitions. While they need not sign up to every single policy, if the population offers popular support for the government, it will encourage the military in following the latter's orders.

Second, your people must support the armed forces. And for this to function effectively, your military should not be perceived as an entity apart from the population—they should be in the public eye, contributing to the public narrative. Senior officers should speak to the media; local units should work together with their local communities, for instance. Conversely, if your military acts in a way that your population deems inappropriate—torturing enemy soldiers for example (not all populations will have the same response to such acts, but most will)—then your population may withdraw their consent.

Your military needs to reflect its public and to be seen to be drawn from all sections of the populace. If, for example, all your military recruits come from one ethnicity, or class, then when you take significant casualties the relationship between your military and populace will come under strain—that group will naturally feel that they are shouldering the burden disproportionately, rather than it being spread across the population of fighting age.

More than this, your military must be seen to reflect the values of your population. If your country is broadly gender equal, then your military should accept women into all roles—to do otherwise risks the military being seen as something apart by the wider population. If your country seeks to end racial and religious discrimination, then your military should encompass all the races and religions in your country. If you are a democracy, then the command style in your military needs to reflect this and have some elements of democratic accountability; autocracies will, unsurprisingly, have more autocratic command styles.

These feelings also work in reverse: the government and military must act, and be seen to act, in the interests of the population. At its most extreme, this means eschewing use of your military against your population, say, to control a riot. But more subtly, the military should take pride in the fact that they are putting themselves in danger, so that the population at large can be safe; and the government must reflect upon the gravity of sending soldiers into danger, neither deploying them flippantly, nor callously. Finland, mentioned above in the context of national mobilisation, is a good example of this three-way balance.

These relationships take decades to build, and minutes to damage. Countries that have strong, long-held traditions of mutual trust between the population, the government and the military can withstand the vicissitudes of an unpopular war (but perhaps not a series of unpopular wars). This relationship must be tended to and nourished over the years, for it is the foundation of good morale and hence the strength of your military.

You should also carefully consider any conflict where you embark on a war of choice and your enemy is fighting an existential war, for example in defence of their country. A people defending their homelands and families will fight with a tenacity and morale that are unmatched. In such circumstances, you must ask your strategic planning teams to consider whether you can achieve your objectives.

This assessment assumes greater significance if your military is conscripted rather than voluntary (we consider the difference in the next chapter). Both the US forces in the Vietnam War, and the Russians in Ukraine learnt this lesson the hard way: when people have their backs against the wall, they will fight with a much greater determination than the conscripts that have been sent to oppose them.

Your military must also win its battles and win its wars. There is nothing more corrosive to this trinity of population, government and military than defeat on the battlefield. Inevitably, the military will blame the government for the disaster—and vice versa. The population will feel let down and have its own views about where responsibility for failure lies. When it comes to morale, defeat must be avoided at all costs.

These elements may come together and lead to unexpected outcomes. Towards the end of the First World War, the Imperial German Navy was ordered to sea to take the fight to the British Royal Navy. It refused, mutinied, and this contributed to the German Revolution of 1918–19 which swept away the old monarchy and instigated the democratic German state, or Weimar Republic. The German Navy's collapse was precipitated by its appalling morale—itself the result of a widespread feeling prevalent in Germany that the government and military were no longer serving the population (who had suffered four years of privation), as well as earlier defeats in the Battle of Jutland and poor rations. The lesson is that your forces' morale keeps the military fighting and your government in power.

Peacetime preparation for battle

Once these foundational ideas have been considered, high morale (and the determination to win it engenders) is developed by progressively more difficult training, high-quality leadership, and firm-but-fair discipline. Suffice it to say, tough training that is appropriate for the types of war that you will be fighting is the single biggest means to improve morale, beyond the factors discussed above.

Your training programme will also develop the leadership skills of your officers and non-commissioned officers. Good leadership is essential for morale, as you will know if you have ever worked for someone who has poor leadership skills. Good leaders share the burdens and risks of soldiering with their troops. They will be next to the point person as they storm the enemy trench; they will help carry the extra ammunition on patrol; and they will reach for the mop and bucket when the toilets need to be cleaned.

Leaders need to be brave, both physically and morally. Physical courage is well understood by all: will they put themselves in physical danger—say, running up to an enemy vehicle to disable it—or shrink away, cowering? It is simply inconceivable that your leaders do not exhibit this physical courage because the bravery of your troops follows on from, and is greatly enhanced by, the valour of your leaders. In the fear generated by battle, soldiers look to their leaders as examples when their natural instincts tend towards self-preservation.

There are innumerable examples of how bravery in battle resulted in a turning point when the momentum shifts from one side to another: for example, in the Falklands War, Col H. Jones, the Commanding Officer of the 2nd Battalion of the Parachute Regiment, personally led a charge uphill against multiple entrenched Argentinian positions. Although killed in the endeavour, he was credited with breaking the morale of the Argentinians and inspiring his troops to complete the attack and occupy enemy ground, changing the outcome of the engagement.

Moral courage is less well understood, but can be summarised as doing the right thing, even or especially when it is difficult or inconvenient. It can be something as simple as owning up when you have made a mistake, or not falling into temptation with a fellow soldier's husband or wife, through to reporting theft or fraud perpetrated by a good friend, or war crimes covered up by your superiors (assuming that your army considers war crimes to be wrong).

The opposite of having moral courage is being corrupt—either by accepting bribes, or in the extreme, skimming your soldiers' pay. There is no faster route to reducing morale than having corrupt leaders—why would anyone charge an enemy machine gun if their leader has been taking a 10 per cent 'tax' on their pay for the last two years? Russia found this out to its cost in the 2022 invasion of Ukraine, as did Afghanistan in 2021 during the Taliban (re)takeover of the country. In both cases, the armies of Russia and Afghanistan suffered extremely low morale, of which corruption amongst the officer classes was a major contributory factor. In the Afghanistan example, the army collapsed in a fortnight.

It is not immediately obvious why moral courage leads to good morale. The truth is that battlefields, and war in general, can appear to be morally ambiguous arenas. Many people will die, often in unfair or innocent circumstances. Leaders often fight wars for one reason, whilst proclaiming other reasons publicly. And, by definition, both sides will have competing narratives or explanations for what is happening. This mix is then interpreted by soldiers who will be feeling extremes of emotion due to their battlefield experiences. In short, in battle, it can be difficult to work out which way is up, and which way is down.

For example, your armed forces have invaded a country based on the narrative that you have given them—that they are liberating the population from a tyrannical government. Yet when your troops get there, they find that half of the population consider you liberators, while the other half regards you as invaders and oppressors. As your troops go about their daily operations, it will be difficult for them to understand what the conflict is about, and this will colour all their interactions—military or otherwise—with the local population. Of course, this 'example' (with differing ratios of opinions amongst the local population) could be drawn from any number of recent wars.

Leadership helps cut through this confusion, because good leaders help their soldiers navigate the moral ambiguity that surrounds battle and conflict. They provide a clear moral framework through which to interpret events, and they can only do so successfully if they too are considered moral and honest. In summary, moral courage supports leadership, and leadership supports good morale.

Leadership and discipline are also closely linked. Discipline is the framework that surrounds your soldiers with expectations of what their behaviour should be, and part of leadership is the policing of this framework to ensure that it is followed. Obviously, a leader cannot enforce discipline if they do not themselves follow it—there is no place for hypocrites in an effective military. Ultimately discipline and leadership keep people fighting and obeying orders even when it appears that the situation is hopeless.

Different armies will have different approaches to discipline: in some, like North Korea, discipline is effectively autocratic tyranny, with executions and soldiers sent to punishment camps for infringements of discipline. At the other end of the scale, most Western armies implement firm-but-fair standards of discipline designed to develop mutual respect between soldiers and officers, and pride in themselves and their comrades. Discipline is meant to act as more of a motivated framework which guides behaviour.

Both leadership and discipline must form part of the culture of your military if you want your military to be successful—and both are built up by the accumulation and repetition of thousands of small acts during peacetime soldiering. Soldiers should be smartly dressed and groomed; they should salute their officers when required; they should arrive on time for tasks; and they should obey legal orders, no matter what they are. Discipline sets patterns about obeying orders, and it creates a framework so people know how to act and it is this—in the moral ambiguity of war—that paradoxically helps them enjoy and benefit from higher standards of morale.

Discipline also has the very practical function of ensuring that soldiers carry out boring, unpleasant and repetitive tasks—for instance digging trenches to the required depth rather than slightly shallower because they are tired; carrying out the required defensive patrols; and camouflaging their encampments properly from air attack—tasks which ultimately save lives.

As you are observing your forces in war, whether they appear disciplined in carrying out these boring or unpleasant tasks is a good way to judge their morale (and hence their effectiveness). All of these behaviours act as early warning indicators for armies with poor morale and may herald significant problems in the future such as desertion, or in extreme cases, mutiny. You should be acutely aware of these signals, and particularly whether they point to the same things as your generals are telling you about your forces' state of morale. Any mismatch not only warns you about the condition of your army, but also that your general is falling foul of the obsequiousness bias.

As should be clear to you by now, the intangible nature of morale is best supported through other intangible elements like a balanced relationship between government, people and military, and through high standards of leadership and discipline. In the absence of these factors, your forces' morale will suffer, and they will be much less effective. But once you have these elements in place, you should consider other, some practical, activities that can support or enhance good morale.

Good morale in war

The greatest contributing factor to good morale in war is victory. Your generals must strive to ensure that your troops achieve early successes in the short term, in order that they might take on bigger, more difficult, and daunting objectives in the longer term. In other words, they should carefully sequence the objectives that they take on, so that your soldiers feel that they have an unstoppable momentum on their side.

Field Marshal Lord Bramall—a British officer who started his career on the beaches of the D-Day campaign and finished as Chief of the Defence Staff in the 1980s—used this technique to great success in Borneo in the 1960s. There, as a battalion commander, he gave his companies steadily more difficult objectives: first an outpost, then a long-range jungle patrol, then an amphibious assault. This sequencing allowed success and good morale to mutually reinforce each other as the conflict progressed. There is no substitute for victory.

Finally, there are other practical steps that you can take to keep morale high. Provide the best medical services that you can, so that your soldiers have the expectation that if they are injured, they might survive. Similarly, do everything possible to recover your soldiers' bodies from the field of battle if they are killed. They must know that their loved ones will be told how they died and where they are buried.

Not collecting and burying the dead, or providing appropriate medical care can even be a problem in non-democratic countries: in recent decades in Russia—during the Afghan and Chechen Wars—committees of mothers have protested and demanded information about their dead and captured offspring. These campaigns create huge political problems for their governments—there is nothing as politically powerful as a grieving mother.

Finally, you must pay your soldiers adequately, feed them properly, and look after their families when they are fighting for you. This is both important in an obvious sense—soldiers want to know that their spouses and children are living well enough. But your people will also demand it for the families of their heroes: this contributes to the foundational relationship between government, population and military.

4.

TRAINING

War—more than anything else—is a human endeavour, and so the quality of your men and women will make the difference between success and failure. Military training extends beyond the individual to create teams, and eventually training creates the teams of teams that make up your military organism.

Training makes an individual more physically fit, more robust, and orders their thinking and actions so that they are socialised to operate within a military team, and within wider military culture. It builds physical, mental and emotional strength. Most importantly, realistic and effective training contributes to good morale, the importance of which you learnt about in the preceding chapter.

You should consider war as the ultimate team sport, and if your teams can remain cohesive amid the brutality of close combat, they will survive and win. And if your teams—for example, the thirty or so people in a platoon—can work together with other teams, then you will be on your way to create the huge, segmented team that is a successful military.

Training is not education. Your training programme should focus on learning skills and behaviours that are practically focussed on outcomes. For example, what do you do if you come under artillery attack? People who have been through the same training programme will respond in the same way when they hear the screech of an incoming shell (by diving to the ground immediately).

On the other hand, education is about acquiring knowledge and is not necessarily practically focussed on outcomes: people who have been taught about artillery will not necessarily respond in the same way to coming under artillery attack (some diving to the ground, some running about, others crouching down and covering their heads, etc.). And while both education and training are needed in a military, this chapter will specifically look at military training.

The basic principle behind your military training should be assembling a team of people and putting them in progressively more difficult scenarios that ever more closely resemble war. This does two things. First, it trains the team, and the individuals therein, how to operate in a technical sense: what are the jobs that they need to do if they are going to manoeuvre their tank across the field, and then destroy the enemy vehicle at the edge of the forest?

Second, through repeatedly facing adversity together in their training, the team will begin to mould into a closer and closer unit, with the closeness often resembling a family. Aside from the obvious benefits of a more cohesive team, such bonding means that the individuals in the team are more likely to put their lives in danger for each other, and in the extreme, sacrifice themselves for their teammates. They will also experience higher morale, which helps them remain cohesive.

This comradeship is one of—if not the—foundation of military strength and effectiveness; it is also what most veterans cite as the thing that they miss the most once they have completed their military service. Get them together in a bar, and they will laugh and joke about all the adversity they have faced—the long marches, the nights without sleep, the high mountains that they have climbed—liberally sprinkled with all the practical jokes that they have played on their comrades when they have been at the limits of their endurance! It is a bond like no other.

* * *

This chapter will set out how to train your military from the individual up to an international coalition of hundreds of thousands. It will explain the different ways of selecting your military personnel and help you choose between different philosophies of training, it will describe basic training, and how long it takes to train highly prized, specialist personnel such as commanders, nuclear submarine engineers, and drone pilots. There then follows an explanation of how to train large formations—like brigades and divisions—and the realities of working in an international coalition (most wars these days are fought with multinational coalitions).

Selection of military personnel

There are two main routes to selecting military personnel—either by conscripting (requiring by law) soldiers to join your military, or through volunteers signing up to serve. Both have advantages and disadvantages depending on what type of force you wish to create.

Conscription is forcing citizens (usually men only, but also sometimes women) to serve in your armed forces for one or two years when they are in early adulthood, usually starting at 18 years old. There are often exemptions for completing education, or for medical reasons. Some countries are relatively relaxed about handing out exemptions, preferring to have a cadre of willing recruits, and in states riven by corruption it is almost always possible to buy your way out of service. This means that these militaries are comprised of poor people whose families cannot afford to do this. At the opposite end of the scale is Eritrea, a very poor country in north-east Africa, where conscription of both men and women is almost total and can in some cases last until the recruits are beyond the age of military effectiveness (usually about 40 for most roles).

Revolutionary France was the first country to introduce conscription, in 1793. At the time, it was fighting several other European countries and was unable to meet its manpower requirements through volunteering alone. All able-bodied men aged 18–25 were required to serve and the army swelled to 1.5 million men within a year. At the beginning of the twenty-first century, eighty-five countries around the world, slightly less than half, have some form of conscription. Many countries that abolished conscription—like Sweden in 2010, and Lithuania in 2008—have reinstated it as nearby military threats—in this case Russia—have re-emerged.

Conscription has the benefits of creating a large military at relatively low cost per soldier. Large armies bring benefits (as Stalin said: 'quantity has a quality all its own') and act as a deterrent, making potential adversaries more cautious about attacking you. They also create a pool of people among your population who possess a basic understanding of military matters, and who can be re-drafted into the military more quickly were an emergency to happen. Finally, conscription can create and enhance a sense of a common national feeling, which is useful for newer countries—Israel, for example, has benefitted in this manner from conscription. Additionally, the population will tend to support the armed forces more because most have experienced military service.

However, conscripted armies tend to be more poorly trained with lower technical skills. Training soldiers takes time, and it is hard to gain enough skills and experience within one year's conscripted service to make a difference on the battlefield. This can limit their usefulness in a conflict, unless you intend to train them to carry out the simplest jobs or are less worried about the casualties that they may sustain given their lower training. Myanmar is a good example of the latter.

Casualties matter little to countries like Eritrea or Russia, but would not be politically acceptable in a rich, democratic country. They wouldn't fare well in China either—its military is a volunteer one—because the one-child policy has led to many families with only one child and a large number of casualties might precipitate serious social unrest. France phased out conscription in 1996 because it realised during the Gulf War in 1990 that it was relying almost completely on its professional, rather than conscript, forces for deployment. Modern warfare is simply too complex for poorly trained soldiers that you want to keep alive.

Volunteer armies tend to be smaller, more highly trained, or technical, and cost more per head. Typically recruits sign up for longer periods—between three and five years—which means that they can be trained and integrated into the military to a greater degree. All other things being equal, volunteer militaries tend to have higher levels of morale and motivation, and this can make them more deployable (or more successful when they are deployed on operations). Conscript armies by contrast tend to have lower morale, by dint of recruits not having a choice in being conscripted, or in what jobs they are tasked to carry out once conscripted.

The size of volunteer militaries—typically around 0.5 per cent of a state's population—means that most citizens will have little experience of the armed forces. This can be a particular problem when a political class has no background in military affairs, as they may not fully understand how to use lethal force to achieve foreign policy goals.

Recruitment into volunteer armies tends to surge when an economy is doing badly, and vice versa. This particularly increases recruitment from among the lower socio-economic classes, and the infantry have long been drawn from the poorest echelons of society. Many of these people transform their lives after signing up—gaining education, income and social standing and making service in the military a powerful agent of social change. Active participation in wars also boosts recruitment. Young men in particular are attracted by the promise of adventure, camaraderie, and the chance to 'prove themselves'. Such feelings often change once they have experienced a war.

As well as the distinction between conscript and volunteer militaries, you should be mindful of the division between your full-time and reserve military. Countries organise their reserves in different ways, but these usually comprise an 'active' reserve—that has to complete a certain amount of training every year—and a 'passive' reserve—which is little more than a list of everyone in the country who has previously served in the military and could be called up in a national emergency.

Reserves offer the option of maintaining some capabilities (for example, specialist engineering or logistical skills) at a much cheaper cost. They also enable you to draw on highly trained individuals from among the civilian population (such as IT experts) that you might have difficulty attracting into the full-time military. Finally, being half civilian, half military is a distinct advantage in civil affairs units that are responsible for liaising with (or providing) civil administration in areas that you control.

A word of caution is merited: reserves give you the impression of having a much larger military than is the case. It takes time to mobilise them and bring their training up to date (three months would be a standard estimate), and there always turn out to be fewer of them than on paper, once those who have unavoidable commitments or medical issues are removed. If you place key war-fighting capabilities in the reserves—particularly logistics—then you must understand that this will limit the deployability of your forces.

The military that most successfully uses reserves is probably the Israeli Defence Force. All Israelis under the age of 40 are eligible for reserve service (once they have completed their conscription). Although, once exemptions are realised, only 25 per cent of those that are eligible serve, with most completing around a month of military training per year—often this is in the same active-duty unit in which the reservist completed their conscription. This model creates a large, trained pool of manpower for Israel to use during military crises.

Philosophies of military training

Before you think about training your forces, you must consider what type of command structure you want. For instance, should everyone blindly follow orders without question even though they will certainly die (because that particular order is a poor order)? In other words, a very hierarchical, dictatorial structure.

Or, at the other extreme, do you want people to question orders, and speak truth to power, so that the decisions that are made are much better options? How much leeway do you want subordinates to have in implementing their orders? What is the role of officers? What about non-commissioned officers like sergeants and corporals? These questions need to be settled before you begin your training programme, because they constitute the basic philosophy of how your military will run, and hence how you will train people.

Militaries tend to reflect the societies that they come from. The most 'democratic' armies are those of the Anglosphere—US, UK, Canada, Australia, New Zealand—and other small European nations like Sweden and the Netherlands who tend to use a delegated command style called Mission Command. Autocratic militaries—where orders are often obeyed without question, even when they may result in mass casualties—tend to be drawn from autocratic societies.

Mission Command works by instructing subordinates in their objectives, but not how to achieve them. So, for instance, one of your battalion commanders will order his company commander to assault an enemy position, but the officer leading the company must come up with his own plan. This command philosophy works—in theory at least—all the way up to the largest formations, and all the way down to sections of eight soldiers. At times, theory and practice may diverge because of the natural tendency—exacerbated by ubiquitous modern communications—of commanders to seek to control the battle, and their subordinates, in detail.

When it works, Mission Command can be extraordinarily powerful because it allows armies to operate in a decentralised way, with commanders at different levels feeling empowered to take the initiative, provided they accomplish the tasks set out by their higher-ranking officer. It means they are more likely to evaluate and take risks; it increases commitment to their part of the battlefield; and it makes it hard for an autocratic enemy to respond to hundreds of micro-decisions and initiatives being made against them. If you decide to adopt a decentralised command philosophy in your military, then you must train for it right from the outset of every soldier's career so that they intuitively understand when and where to take the initiative, and when and where to follow orders blindly.

Linked to the question of command philosophy is that of the role of officers (Lieutenants, Captains, Majors, etc.) and non-commissioned officers (NCOs) promoted from the soldiery (Corporals, Sergeants, Warrant Officers, etc.).

Professional, volunteer armies from democracies tend to have strong layers of NCOs sitting between the soldiery and the officer class. The NCOs are the middle management of your military and because they serve a long time in comparison to soldiers and most officers, they become repositories of experience. In these mostly Western armies, officers should be seen as the generators of plans and ideas that link lethal force to political intent, and the providers of overall command and leadership, whereas NCOs can be seen more as the executors and leaders of those plans at the small unit level (platoon and company). When it works, this distinction can be a very powerful combination of command and control, not to mention initiative.

Conversely, autocratic conscript armies like that of North Korea tend not to have empowered cadres of NCOs operating with initiative, and the job of all layers of command is seen as passing down orders from above. This can result in an inflexible military unable to respond to setbacks or changes in the situation.

The advantages and disadvantages of the two styles were clearly demonstrated in Ukraine, where poorly-trained Russian conscripts were ordered into frontal assaults more reminiscent of the Second World War than the twenty-first century. On the other hand, the Ukrainian forces, who had benefitted from eight years of training in a more decentralised command structure, operated in much smaller units who could seize the initiative and attack Russian units from the side and the rear. As a result, Russia suffered extremely high casualties which, at the time of writing, had reached potentially up to a third of their initial deployed force.

Training the individual soldier

Basic training is the cornerstone of your training programme. Although each armed service—navy, army, air force—has its own programmes, all military basic training seeks to achieve the same things: improved personal skills, inculcating the ability to work in small teams, socialisation into the wider military system, and some basic military knowledge (usually in that order). In most professional militaries basic training takes around ten to fourteen weeks. This section will focus on the training you should give to army recruits, but much of it is common to navies and air forces as well.

Training is about getting individual soldiers to work together as a team, and to keep operating as a team amid the shock and brutality of close combat, while maintaining high morale. It also helps individuals and teams carry on instinctively when they haven't slept for two days, or when several of their friends have been wounded or killed. Throughout their careers, soldiers should carry out approximately 25 per cent of their training at night—most armies don't reach this amount of night training—and it will offer you a distinct advantage on the battlefield.

You should design your training programme around maintaining two principles: if individuals or teams are unable to look after themselves, they automatically become a burden to others; and, the harder your training is, the better prepared your soldiers will be for combat. Field Marshal Bramall, whom we met in the last chapter, put it like this: 'Officers who believe their soldiers will be grateful for soft and easy training do not understand human nature and will never succeed in producing high morale.' It was put more pithily by a Russian general in the 1700s, Alexander Suvorov, who said: 'train hard, fight easy'.

In making the transition from civilian to soldier and keeping in mind that many of your military recruits will not have benefitted from a good education, basic training seeks to improve the personal qualities of individual soldiers. Physical fitness is obviously important, but training should also be designed to improve self-confidence and discipline, physical courage, and self-control. Self-control in particular lays the foundations for the controlled aggression that all soldiers need to switch on and off when they must kill people. Controlling the level of aggression that you deliver to the enemy is as essential in the individual infantry soldier as it is for the general, who is also using violence to communicate, albeit at a much higher level.

Finally, basic training is the start of the reinforcement of moral courage and integrity—doing the right thing—which is considered a key concept in most professional military forces. Your military will have to operate far from home, fighting enemies who may act in morally ambiguous ways, yet you will need your forces to continue operating legally and with integrity. Many armies fail in this regard and are less effective as a result—committing war crimes or atrocities usually hinders the achievement of your foreign policy goals. Soldiers operating with integrity also encourage trust between one another, which is the foundation for teamwork and high morale. Emphasise the moral character of your forces in your training programme.

Your training must also encourage the extremely accomplished teamwork that is characteristic of all successful militaries (and is quite different from the levels of teamwork required in civilian life). It starts with individuals working in pairs, then groups of four, and then eight, and so on. These teams are then put in challenging circumstances, forcing them to work together, trust each other, and rely on each other. Many soldiers make friends for life in their basic training.

Mutual trust and support are essential on a battlefield where, for instance, one soldier must sleep safe in the knowledge that their comrade on sentry duty is awake and listening for the approach of the enemy. Ultimately, if done properly, these teams should be full of individuals who put the team (and the mission) before themselves. It is also one of the reasons that militaries have a predilection for marching drills—because they psychologically prime individuals to be aware of others, the better to act in synergy with them.

Third, basic training is about standardisation, so that recruits are socialised into the wider military system. This is particularly important because it makes large militaries flexible in that they can take individuals, teams and small units, put them together at short notice and in chaotic situations, and expect them to operate efficiently. This standardisation crosses the breadth of military service—from language (all militaries have their own internal languages, slangs and humours), to the way teams and organisations are structured, and from symbols flags and badges, to using the same military equipment—a rifle, for instance—in the same manner.

Finally, your basic military training should cover a series of simple military skills. These will include rifle usage and marksmanship (how to shoot accurately at ranges out to 300m), map reading and navigation, the use of kit and weapons that all soldiers carry, and simple, small-team-level military tactics—how to storm a trench, how to manoeuvre around and onto an enemy position, and how to use a bayonet on your rifle to close with and kill enemy soldiers. (Bayonet training is also a key part of imparting controlled aggression because soldiers are brought into a heightened state of emotion in order to kill, and then brought back down again during the exercise.)

Even if an individual soldier is a specialist who works miles from the front line, everybody needs this basic training—in a war, you will never know who will need to go where. These skills provide a foundation for more technical training—like firing artillery or driving a tank—but more importantly, everyone must be familiar with the most basic, brutal forms of warfare at the small-team-level because turning-points in conflicts tend to be settled by groups of infantry holding or taking critical positions—bridges or hills, for example—on the battlefield.

It is not only soldiers that receive this basic training, but airmen and sailors too, with slight variations in emphasis. All military personnel must execute basic tasks automatically in concert with their comrades. If they are unable to do so, they become a burden, hindering the effectiveness of that unit yet further.

Specialist training

You must also develop training specific to the various branches of your armed services. For the air force you will need to develop pilots, ground crew, and maintenance personnel. The navy requires engineers and weapons officers and the army engineers and tank crews. All three services must have intelligence specialists and logisticians. Different specialisms take different amounts of time to train—it might take up to four years to train a jet pilot.

The longer it takes to train somebody, the harder they are to replace, and the higher priority a target they become on the battlefield for your enemies. At the extreme end, a general might need twenty-five years of training and experience to successfully manage operations on a complex battlefield, making them an extremely high value target. This was demonstrated in the Russo-Ukraine War, where Ukraine successfully targeted many Russian generals, severely hampering Moscow's ability to coordinate its forces (we will look more at how to do this type of targeting in Chapter 9).

The principles of specialist training are much the same as basic training: teach your soldiers, sailors and airmen technical skills—how to drive a tank, for instance—and then focus on developing those skills with other people in small teams—getting the tank driver, gunner and commander working together so well that they function as a quasi-organism.

There follow a few examples of specialist training, to demonstrate to you how difficult it is to generate some of the specialist capabilities needed on the modern battlefield—and how crippling it can be when you lose these trained people to enemy action.

Let's look at an infantry solider. Once a recruit has undergone basic training, they are far from being a useful, combat-ready infantry soldier. You should plan for an infantry course of about six months, to bring your soldiers up to the standard required—

this would cover light role infantry, and armoured infantry would require further training. The course should cover the innumerable tactics required to survive and fight as companies and battalions of infantry, including how to coordinate an attack in formation, how to manoeuvre across the battlefield without being killed, how to communicate effectively and securely, how to storm an enemy-held position, how to flank the enemy, how to retreat in good order, how to defend an urban area, etc.

All these tactics require an ability to read the ground—that is, to assess small rises, hidden dips and other topographical and hydrological features and how they will affect you and the enemy. This is the real art of the infantry soldier and although it can take years to develop full expertise, the basics must be inculcated in your specialist infantry training course.

Beyond this, you must train infantry soldiers on a wider range of weapons than simply the rifle—all light role infantry formations will use heavier machine guns, mortars, and anti-tank weapons. Your soldiers will need to be trained in advanced first aid, and how to live in the field for an indefinite amount of time. And finally, as mentioned above, your soldiers will do lots of marching and drill, helping them psychologically to operate in close synchrony with their comrades as they execute the infantry battlefield tactics that they have learnt.

The training for your tank crews should also be around six months. Individual recruits will either be trained as tank drivers, gunners or loaders (the fourth role in a tank—the commander—will only be given to someone who has a few years of experience or is an officer). The driver, gunner or loader training will take about six weeks of the course and the remainder will be spent learning how to communicate, how to maintain their vehicles, how to administer first aid, and the basic tactics of operating with other tanks on the battlefield. But as with the infantry—the most important output of tank crew training is that they all know instinctively how to work seamlessly with the other members of their crew in the heat, chaos, and cramped conditions of battle.

These training programmes only get longer for your more specialist troops. Take military engineers for instance. Although you can train them in about ten weeks in the basics of combat engineering—clearing mines, constructing bridges, crossing water obstacles and blowing things up—it can take up to a year to train a metalworker, or a building surveyor, or a cartographer.

If you intend to operate in countries with different languages from your own you will need linguists and depending on how different their language is from yours, it may take up to eighteen months to train them. Commando forces might take a year or eighteen months. Probably the longest training programme of all in your army would be for a helicopter pilot—which can last for up to three years. These simple mathematical facts will make your helicopter pilots high priority targets on the battlefield.

In your army, your mindset should be that of equipping your personnel; but in your navy and air force you should think about manning your equipment—the scale and size of your equipment will be larger and more complicated in the latter two services. This will lead to very long training pipelines to produce key personnel.

The training for a Principal Warfare Officer (PWO) in the Royal Navy—in charge of the operations of a ship—takes around a year, on top of their basic training, their officer training, and several years of more junior experience. They must know how to command and manoeuvre the vessel, be expert in naval tactics, operate and command all the weapons systems on the ship, conduct anti-submarine and anti-air operations, and know how to operate the ship as part of a flotilla, or in support of land operations. In short, it is an extraordinarily complicated job.

Air force fighter pilots must first undergo basic training and officer training. Then theoretical flying school, and basic flight training on a simple propellor aircraft. Then they learn how to fly jets, learn aerial tactics, before training on frontline operational jets, and learning all the weapons and tactics required to operate either strategically, or in support of ground forces. All this training is interspersed with pilot specific instruction, such as survival and escape and evasion training so they know what to do if they are shot down in enemy territory.

Finally, you must consider that the personnel with the longest training (and experience) burden are commanders of military formations. A Major—in charge of a company of around 100 personnel—will take nine years to reach that rank; a Lieutenant Colonel—in charge of a battalion of 500–1,000 personnel—may take sixteen years to reach that rank; and a Brigadier—heading a brigade of, say, 5,000 personnel—will take around twenty-four years to complete the training and gain the experience necessary to make them effective in their posts (the rough rule of thumb is that there are three companies in a battalion, three battalions in a brigade, and three brigades in a division).

In major wars, these timescales inevitably become shortened—Roland Bradford, a British officer on the Western Front in the First World War, was promoted to Brigadier after merely five years in the regular army—but this is obviously less than ideal (he was also killed shortly after his promotion).

Formation training

You will be aware by this point that the military is a team of teams. You know how long it takes to create teams at the smallest levels. Training combined arms land formations—that is, brigades (3–5,000), divisions (10,000+) and corps (30,000+) where infantry, tanks, artillery, and other military specialists work together—is a continuous process called 'force generation'. This is particularly true when individuals will be posted in and out of different parts of your military, and others will be leaving service.

You must allocate sufficient funds for your military to conduct this continuous training, particularly the costs of fuel and ammunition (higher formation force generation is often one of the first things that is cut from military budgets because it is so expensive). Without conducting training at this scale, commanders and teams will not experience the frictions that exist at this level of war-fighting—how limiting logistics can be, how crippling the loss of a communications node can be, how exactly to integrate aviation into your ground manoeuvring, or how the loss of capability of one of the triad of artillery, armoured and infantry can unbalance your force, making it highly vulnerable.

On this scale, war is an orchestra that must be conducted, and without continuous practice, commanders and teams simply will not know how to do it. Very few countries can afford exercising in such huge numbers: the US and French armies still do it; the British have not done so for some time due to financial constraints. The Russians have appeared to conduct these levels of drills in recent years but judging by their poor performance in Ukraine, their training was not terribly effective, or simply done to demonstrate Russian capabilities to its would-be adversaries rather than being of concrete value.

These so-called 'force generation' cycles should be developed over two or three years. The first year should comprise individual training and courses for people to learn specialist skills, such as how to guide an aircraft onto a target in support of ground forces. It is also during this first year that you should integrate any new equipment, train your people on it, and iron out short-term problems.

The second year should be devoted to progressively larger cycles of training—starting at platoon, then working up through company and battalion level. At the final stage different types of military capability will start to integrate with each other in what is called combined arms training. Thus, commanders at the battalion level—now called battle groups if they are combined arms units of different specialisms—will learn how to manoeuvre infantry, tanks and artillery together with reconnaissance forces. Other capabilities, such as engineers, will be added to the mix. Combined arms training is a big shift from training only within your own speciality, and professional militaries spend a great deal of time doing it.

Beyond battlegroup level, your troops will need to train in brigades, divisions and corps. It is at these levels that all the military capabilities under your command must be integrated and trained together—most importantly, your logistics elements that only exist at brigade level and above, but also helicopters, specialist artillery, like rocket systems, air and naval support, and other newer capabilities like cyber operations.

The other important reason for training at this level is so that your headquarters staff learn how to direct war-fighting at this scale. They will form a cadre of 'staff officers' whose job it is to provide intellectual and administrative support to the commander of that formation—and they must be trained and have gained experience just as the troops in the field.

This type of training is extraordinarily expensive, costing millions of pounds at a time. You need a large area of land—the British Army use training areas in Canada—to which you must transport your armoured division of several hundred vehicles and thousands of people. To this must be added the logistics, fuel, ammunition and spare parts. Then you need an opposing force, for them to 'attack', and an entire infrastructure to run the exercise over three months to generate as close as possible the challenges that you will experience in real combat.

If your army does not train at this scale, then it will be unable to deploy at this scale. That is, if you have a corps' worth of troops and capabilities, but only train at brigade level, when you deploy on operations you will not be deploying a corps, but nine separate brigades, who will find it far harder to work together. This will make it much easier for a moderately competent enemy to separate and destroy them individually.

Training in coalitions of allies

Finally, you must consider how to work with allies: many modern wars are fought in coalitions made up of troops from different nations. This is largely for two reasons. First, during the twentieth century, war has become a less and less acceptable way of settling differences between countries, compared to previously, when it was considered part of the natural order of things. Hence it is easier politically to claim an overriding justification for a war—say, to depose a brutal dictator or to degrade a terrorist group's infrastructure—if several nations work together to prosecute a war. Second, creating a military force able to fight a modern war is an incredibly time-consuming and costly endeavour, and if you can fight alongside other countries, some of this burden can be shared.

The two reasons tend to interact where a large powerful state prefers to fight with the political support of as many other countries as possible. The former will provide the overarching command structure for the coalition and handle most of the logistic support, particularly in transporting war materiel to the theatre of operations. In return, the smaller countries, who have also decided it is in their interests to fight this war and offer political backing to the powerful nation, can provide smaller packets of forces that slot into the bigger coalition. In essence, political and logistical support are traded.

Although they can bring benefits of military mass and political cover for your war, working in coalitions is not for the faint-hearted. There are the problems of friction between the forces of the various nations fighting together—all countries have differing military cultures. Moreover, forces are usually embedded in a shared command structure—that is, your nation's forces sit under the operational command of another nation's forces—with political caveats as to what those forces can and cannot do.

For instance, there were more than one hundred political caveats for use of force—in practice red cards—covering the thirty-six NATO countries on the ground in Afghanistan. These ranged from not being allowed to operate at night, or to go beyond the capital, Kabul, or even—amazingly—to take part in combat. These caveats exist because even in a military coalition, component nations may not share the same strategic interests and will have different approaches to the risks of fighting a war.

You may decide that the only way to achieve your foreign policy and military goals is to work within a coalition. If so, then you must prepare for it several years in advance. In addition to standardising all your equipment, ammunition and supplies, you will need to train together regularly—at the same frequency as the force generation cycle outlined above. Some coalitions—NATO, for instance—have standing multinational headquarters which have units from the different nations assigned to them, so that they can train continuously together. You must also work to harmonise your political and economic goals with those of your allies and other countries, to remove the need for caveats on the use of military force within the coalition.

* * *

You have now considered how best to form strategy, deliver logistics, create and maintain morale and train your forces. These are the four foundations that will enable you to project military power. In Part 2, you will learn about different military capabilities—from land, sea and air, to cyber and nuclear—and what they can and cannot do. And in Part 3, you will learn how all these elements—the four foundations, and the four sets of capabilities—come together as you learn the art of using violence to achieve your goals.

Part 2

TANGIBLE CAPABILITIES

5.

LAND

In Part 1, you learnt about the foundational elements of warfare: strategy, logistics, morale and training. Without getting these four elements right, you will not have learnt how to fight a war. Nor would there be any point in throwing your troops into battle. You will simply waste them and your non-human resources.

Part 2 will cover the different arenas—militaries tend to call them domains—in which war is fought: on land, at sea and in air/space, in the information and cyber domain, and finally the altogether different category of fighting with nuclear, biological and chemical weapons.

The most important lesson here is the primacy of the land domain. No matter what anyone tells you—and there will be plenty of sailors, airmen, and (especially) evangelists for new technologies that will try to convince you otherwise—the land domain is pre-eminent, because wars are won or lost only on land. If you only take one lesson away from Part 2, make it this one.

The primacy of the land domain is straightforward to convey. People live on land, and war is a human phenomenon driven by the most powerful of emotions. The reality of trying to influence them is that throughout history wars have always been decided by your troops taking control of someone else's village, town or city and, bearing a sword, musket or rifle, imposing your order.

The other domains that we will cover in Part 2—sea, air & space, cyber & information, and weapons of mass destruction—exist to support the land domain and your land forces. You cannot win a war without them. Nor could you win a war by relying only on the other domains.

On several occasions throughout history, leaders have decided—or allowed themselves to be convinced—that land forces are not needed to win wars. And it is generally new, untested technologies that seduced them to make what ultimately turned out to be a hubristic decision.

For example, in the 1920s, when air power was a relatively untested concept, and because bombing was cheaper than maintaining troops, Britain flirted with using only air power to put down rebellions in her empire. Initially successful in places like Somaliland, Iraq, Palestine and the North-West Frontier of British India (now in Pakistan), by the end of the 1920s, what was known as 'Air Policing' was found wanting: when real resistance or complex political problems were met, ground forces always had to be sent in. There were echoes of the 1920s British Iraq policy in the 1990s US/UK Iraq policy which also attempted to control Iraq from the air via the establishment of no-fly zones and the bombing of Iraqi targets. This was largely unsuccessful, and the two powers decided on a land invasion of Iraq in 2003.

Similarly, in the first decades of the third millennium, Western countries flirted with the idea that wars could be won in the cyber and information domains—that is, over computer networks rather than by deploying tanks. In an echo of the development of air power in the 1920s, fighting in the cyber and information domains is a lot cheaper than maintaining and exercising armoured divisions. In peacetime, cost is a beguiling driver of military strategy for leaders, but it leads to the key fallacy highlighted in the introduction: that technology is the solution that solves our problems in warfare.

For example, in 2021, the UK published one of its periodic defence assessments, *The Integrated Review*. Notable in this review was the reduction of investment in ground forces compared to new investments in cyber warfare, artificial intelligence, space, and directed energy weapons. The army was to lose troops and armoured vehicles, including tanks. Brilliantly illustrating the fallacy outlined above, the Prime Minister at the time, Boris Johnson, exclaimed that 'we have to recognise that the old concepts of fighting big tank battles on European land mass are over, and there are other, better things we should be investing in'. In February 2022, a little over three months later, Russia invaded Ukraine with 200,000 troops arranged largely into armoured formations.

I hope by now you are convinced that land warfare is critical. This chapter will describe how to build and deploy your land forces. It will cover the timeless practicalities of climate, weather, terrain, topography (the natural and artificial physical features of an area), hydrography (water bodies) and towns and cities. It will then consider the different types of land forces at your disposal—their role, their strengths and their weaknesses. Finally, the chapter will demonstrate how to combine these elements into larger and larger formations to create an army.

Climate and weather

Warfare has historically been a seasonal activity conducted during the warmer summer months when food was more widely available and terrain dried out and became passable. The Romans counted on a fighting season running from mid-March to mid-October, much the same as obtains in Afghanistan today.

This season could be even further shortened to the period between planting and harvesting—about three or four months—during periods of labour shortages where men were needed for agricultural work. It was not uncommon for fighting to stop to bring the crops in, and for it to continue afterwards. This is obviously less relevant in areas of the world that do not have such marked seasons, like the tropics.

Modern armies are not so restricted. With very few exceptions they bring their own food with them (which can be stored inter-seasonally), and do not live off the land that they occupy. They also benefit from modern technology which enables them to mitigate the effects of weather on the passableness of terrain. Again in Afghanistan, the NATO armies were largely unaffected by the winter and could operate year-round. It was the Taliban that came out to fight only during the summer!

In some cases the climate is so severe that even modern armies are incapable of operating at scale. (Specialist troops like special forces or commandos can operate in all climates, but they are trained and equipped only to carry out limited objectives.) The most obvious examples of the limiting factors of climate come from the invasions of Russia. In 1812, Napoleon even got as far as capturing Moscow. But he started too late—in June—and up to two-thirds of his men died of disease or froze to death.

One hundred and thirty years later, Hitler also launched his offensive into Russia in June. Despite this being the largest invasion in history—comprising 3.8 million troops—the German advance was beginning to slow by October due to bad weather. By December, the Wehrmacht had reached the outskirts of Moscow—the furthest east that they advanced, losing 830,000 men in the attempt—but by then the snow blizzards had started. The Germans were ill-equipped for the Russian winter and the Russians launched successful counter-attacks, thus marking one of the major turning points of the Second World War.

The lessons are clear: you should always consider the season in which you launch a ground campaign. Even if it is not stopped dead by the winter, as these two campaigns were, you will still find it easier to move your troops, and most importantly your supplies, in summer rather than winter. In the tropics, you will prefer to operate in the dry rather than the wet season: both for mobility, but also to minimise contracting diseases (which are more prevalent when it rains). Your vehicles will also be affected by the different seasons, for example if the temperature is very high, or you are operating at altitude, or both, your helicopters will have a significantly reduced range because of the thinness of the air.

Once you have decided in which season to launch your operation, the daily cycles of the weather will affect how you fight. If very overcast or raining, most of your satellites will be unable to observe the land; your satellite communications may even be affected in particularly severe conditions. In overcast weather, or where the cloud base is low or shrouding the land, aircraft cannot support your ground troops, and helicopters may be grounded. Generally, poor weather favours defenders over attackers because the ground quickly turns into a quagmire. These factors may persuade you to postpone operations until the weather improves.

The phases of the moon will also affect your operational design. You may wish to use the veil of night to move supplies, or to launch attacks: in which case you will want no moonlight. Conversely, you may want the advantages of night, but with light so that your troops can see what they are doing. For amphibious operations, the state of the tide is key—landing at high tide allows your landing craft to get higher up the beach but doing so at low tide enables you to clear obstacles while they are uncovered. If you are planning a particularly complex operation, all these things will be considered.

For the 1944 D-Day amphibious assaults in Normandy, the first criterion was that they had to take place in summer—this was because of the likelihood of a calmer sea state, and a sufficiently early dawn. Next, low tide had to fall around this dawn, because it meant that the defensive obstacles erected by the Germans would be uncovered. But the tide must be rising, so that the landing craft would not get beached once they had unloaded their troops. Finally, there had to be a full moon the night before so that airborne insertions of parachute and glider troops could occur safely. This left a very small number of days that were suitable, and in the event, everything had to be delayed by 24 hours because of a storm.

Terrain: topography, hydrography and urban areas

As with climate, when considering terrain you must start at the largest, continental, scale and work your way to the smallest. You are looking for a route from you to your enemy through which you can get your forces and—as we learnt in Chapter 2—move supplies in any season. This means taking account of topography (mountains, valleys and plains) first, and then hydrography (rivers, lakes and marshes) second. In some cases, you will also have to consider urban areas.

In the case of high mountain ranges, modern armies with their armoured vehicles and huge logistics tails are restricted as much, or even more than, armies were 2,000 years ago. Mountains are, and always have been, simple hard facts of geography when it comes to land operations.

The topography of Europe and western Russia demonstrates this well (see Fig. 3). There is one main route between France and Germany through the Low Countries (avoiding the mountainous Alps), and one eastward from Germany towards Russia through Poland (avoiding the Carpathian Mountains). This latter route then splits into three: through the Baltic countries, through Belarus, and through Ukraine. Wars between great powers in Europe have been fought through these corridors since antiquity, much to the misery of the people who live there.

Now consider rivers: you will see that as you advance along these corridors you will come up against riverine features, with crossing points often marked by cities. Marshland and swamp-land too will force you onto certain routes—they are fatal to everything except light infantry.

In this example, if you decide to take the northern route between Germany and Russia via the Baltics, then the River Vistula in Poland is a major barrier that you must cross. This means securing riverine crossing points in the cities of Warsaw, Krakow or Bydgoszcz (which is one of the reasons why those cities were founded there in the first place). If you take the southern route through Ukraine, you will need to cross the River Dnipro, which will mean securing the cities of Kyiv, Dnipro and Zaporizhzhia.

When it comes to rivers and marshland, modern technology does help somewhat more than with mountain ranges. Your engineers may be able to deploy bridges across narrower rivers or build roads through marshland. And while the ancients could build bridges and roads, they could not do it as quickly as modern military engineers who can deploy temporary military bridges from vehicles or use mechanical diggers to build levees (raised areas or ridges of ground).

However, if your enemy is defending on the line of the river, you must be aware that opposed river crossings are very difficult, particularly if your enemy has artillery or air power still available to them. Similarly, in marshland, your engineers may be able to build or improve existing roads to enable your logistics to pass through, but this will not help if your tanks are required to spearhead the advance.

Now consider the position between China and Russia (see Fig. 4). The two countries share two stretches of border separated by Mongolia. The shorter western stretch is mountainous and completely unsuitable for a large land invasion. This leaves the eastern stretch which runs across a flat plain—much more suitable for large-scale movement of troops. The border itself is largely demarcated by the course of River Amur so any invading force would first need to secure several crossing points. As an aside—consider the finger of Russian land that stretches towards the city of Vladivostok. In military terms, the city would be very vulnerable if China decided to extend the border along the Amur River all the way to the Pacific Ocean. This would then place the Russian island of Sakhalin in a very precarious position.

These issues mean that often there is no option but to take control of enemy cities: not only are they often crossing points of rivers, or gaps between mountain ranges, but often cities are symbols of power, economic centres of gravity, and nodes of government administration.

Urban areas are extremely easy to defend, and extremely difficult to assault. This is because of their complexity in three dimensions—buildings rise out of the ground, and subterranean spaces extend well below in the form of car parks, metro lines and cellars. This three-dimensionality creates innumerable opportunities for reconnaissance (e.g., mortar or sniper spotters in tall buildings), concealed routes of advance (e.g., underground metro lines), safe storage of supplies (e.g., underground car parks), or of troops from bombardment (e.g., cellars) and ambushes from multiple directions (e.g., trying to cross a city square).

In addition to this extreme complexity is the question of civilians: cities are very densely populated and if your army cares about minimising civilian casualties operating in them is extremely difficult. Conversely, if your army cares nothing for enemy civilian casualties then laying siege to an enemy city may solve your 'military' problem but may inspire your enemy to fanatical resistance elsewhere because you have signalled to them that they will die anyway so they might as well resist (or attack!) until death.

We can see from these simple examples that climate and terrain are extremely important in framing when and where you should conduct your land operations: you will likely end up with a small number of possible routes through which to advance your military operations.

The different types of troops in your army

As you will have realised by now, your force needs to draw upon different types of troops. In a modern army, your infantry forms a backbone suitable for all types of operation from attacking to defending, and operations other than total war, like counterinsurgency.

Infantry usually comes in three configurations: light role, mechanised, and armoured. Light role infantry is traditional infantry, on foot, and little changed in thousands of years except in terms of their equipment and weapons. Mechanised infantry is highly mobile and moves around the battlefield in lightly armoured infantry fighting vehicles thus keeping the troops inside safe from most types of attack until they are needed. Armoured infantry traverse the battlefield in well-armoured vehicles, often bearing large calibre weapons, and can withstand much heavier attack.

When advancing or attacking, armoured or mechanised infantry pair with tanks and artillery to make a highly effective triad. These three capabilities are usually grouped together in 'combined arms' formations of battle groups, brigades and divisions. This is because each one has strengths and weaknesses, yet when combined in the hands of a skilful commander the weaknesses can be made to cancel each other out, in the manner of the game of rock, paper, scissors. Let's go through each capability in turn.

Only infantry—of whatever sort—can assault enemy positions that are held by enemy infantry dug into trenches in the ground (known as 'dug in'). Only infantry can clear a trench or assault a building. Similarly, only your infantry can hold or defend territory, so much the better if they are dug in.

The huge advantage of infantry is that when on foot they can operate across all terrain including the urban environment. But infantry's great weakness is that, once out of their vehicles, they are highly vulnerable to artillery, tanks or any other type of attack. They usually have short-range weapons of approximately a kilometre with the exception being where the infantry is armed with anti-tank (sometimes known as anti-armour) weapons: missiles that can destroy armoured vehicles at a medium range of about 4km.

You may also decide to develop some specialist infantry that is highly trained for a narrow range of tasks—sometimes known as commandos, or special forces. These often capture the public imagination with their tales of daring-do, but in reality are very small in number, lightly equipped, and only suitable for completing a narrow range of reconnaissance, sabotage, and assassination tasks (which they do very well). They should never be substituted for regular infantry, who rely on much larger numbers and heavier equipment to complete a broader range of tasks.

Your tanks—still called the cavalry from when they were horses instead of vehicles—are very well protected with armour and can destroy other armoured vehicles up to 3km away with their cannons. They can also kill enemy infantry (providing they don't have anti-tank weapons, to which they are highly vulnerable) and destroy their artillery, aided by their speed of advance. Tanks manoeuvre across the battlefield rapidly, provided it is rolling countryside—comprised of fields, light vegetation such as hedgerows rather than trees—and small watercourses.

Their weight precludes their use in swamp and marshland, and mountains and wider rivers will impede their movement.

Finally, they are particularly vulnerable when attacking in urban environments because defenders can close upon them unseen, allowing grenades and improvised explosives to be thrown onto the tank or (more effectively) through its hatches.

Artillery makes up the third leg of the triad. It delivers munitions at very long range: approximately 15km for 'traditional' howitzer-type artillery, or up to 500km for some modern rocket artillery systems. The latter have the additional benefit of being GPS-enabled, thus being very accurate (traditional artillery is an 'area' weapon, that scatters shrapnel over a wide area, rather than a pinpoint one).

Artillery can fire over hills and other obstacles, directing fire on an enemy that they cannot return, unless they have their own artillery of similar range (tanks and infantry, with the exception of mortars, can only fire directly at an enemy that they can see). With the right munitions, it can be useful against armoured vehicles: a direct hit will destroy the vehicle, but even a near miss will strip the antennas and other sensitive equipment off the tank making it combat ineffective. Also coming under the classification of artillery are air defence missile systems, whose job it is to shoot down enemy planes and drones that attack your forces.

Artillery is supremely effective against infantry that is not dug in (i.e. in the open) and supply dumps (which tend to sprawl over a large area without protection). It has similar effects on civilians, unfortunately. Of the 20 million military and civilian deaths in the First World War, 12 million of them were caused by artillery.

Artillery is, however, extremely vulnerable to attack of any sort, and extremely supply intensive: you will remember from the logistics chapter that it does not take long to fire a shipping container full of ammunition from a battery of artillery.

Artillery is also a more complicated organism than tanks: for it to work effectively, you need an observation team (although these days this is often done by drone), another team working the guns, another team surveying new gun positions, and another team working on logistics. The destruction of any one of these teams can make your artillery ineffective. Artillery—unless it has tracks like tanks do—is limited to roads.

You will appreciate from these brief descriptions of infantry, tanks and artillery that combined they will give your commanders the ability to choose from weapons range, manoeuvre, lethality (literally a weapon's ability to kill), all-terrain capability, and defensive ability. Each leg of the triad also covers for the weaknesses of the other: infantry protects your tanks and artillery at close range; tanks enable you to manoeuvre quickly; artillery enables you to project at long range; infantry enables you to defend ground; tanks enable you to assault hard; artillery enables you to interdict enemy supply and provides cover for your advance, and so on.

You must ensure when creating your combined arms formations that you have appropriate balance between the three types of troops, otherwise you risk jeopardising your entire force once your enemy identifies and exploits your weakest spot. Too little infantry? They will attack your tanks with anti-tank missiles. Too few tanks? They will manoeuvre around you. Too little artillery? Your forces will be pinned down by the enemy's artillery and held off at a distance.

Once you have got the balance right between the three capabilities, you should add reconnaissance troops that advance ahead of your main body of infantry, tanks and artillery. Their job is to find the enemy, assess their capabilities, and investigate terrain features or infrastructure. Traditionally reconnaissance troops were mounted on horseback; in the twentieth century, on motorbike, 4x4, or lightly armoured tank; nowadays reconnaissance may often be done by micro drone (we will look at drones in the next chapter). These four capabilities—infantry, tanks, artillery and reconnaissance—are your fighting troops.

These fighting troops are supported by a number of specialist troops, the most important of which (after logistics) are your military engineers. Their responsibility is to ensure that you can move around the battlefield (known as mobility), stop the enemy moving around the battlefield (known as counter-mobility), and build your defensive structures where they are needed (known as survivability). Because they are so essential to your fighting troops, engineers will be embedded down to battlegroup (battalion) level.

When ensuring your battlefield mobility, engineers will have several tasks. At the lowest level, they will be helping your troops move around by blowing up obstacles in their way—be it putting holes in walls, felling large trees that are obstructing your tanks, or blowing up enemy obstacles like tank traps. They must also build or repair roads and fill in ditches that are too wide for your vehicles to cross (most armies have an armoured tractor/digger for this). Your engineers will also have to clear enemy minefields, a daunting task.

Most importantly, your engineers must know how to build sturdy bridges that can take the weight of your main battle tanks (c. 70 tonnes) across water features of varying widths. If you don't have it already, you should invest in bridging gear, which is a bridge that comes on the back of a lorry, and either folds out in a few minutes across smaller rivers or, for larger rivers, comprises a flat-pack bridge that can be assembled in a matter of hours. Be warned—these bridging assets will become one of the highest priority targets for your enemy, and vice versa. Without them, you will be unable to advance your army across rivers without taking highly-defendable urban areas.

When it comes to counter-mobility, namely, stopping your enemy moving, it is much the reverse of ensuring your own mobility. Your engineers will be tasked with blowing up bridges that stop your enemy from crossing water features. The Armenian army blew up bridges once it had retreated over them in its 2020 war with Azerbaijan, thus enabling it to control the pace of retreat.

Your engineers will also be instructed to destroy roads and runways, usually by cratering them with explosives, and digging wide anti-tank ditches. The purpose of all of these activities is either to stop your enemy advancing, or to restrict their choices and force them to advance where you would like them to: perhaps through a highly-defendable city, or perhaps through an area where you hold the higher ground and can direct artillery fire onto them. This is known as channelling: forcing your enemy to go down particular channels where they can be ambushed or attacked.

Lastly, your engineers must ensure the survivability of your forces by building bunkers for your command posts, overhead cover for your troops (to protect them from artillery), and buildings for your supply depots. They also should know how to construct deceptive decoys like fake buildings and other structures that from the air look like supply depots or headquarters. Much better that the enemy attack these, rather than the real things!

To combat engineers, you must add engineers and mechanics for all your equipment—you will need expert vehicle and generator mechanics, specialist electrical engineers, and innumerable other specialities to keep your army on the road. In addition, your vehicle mechanics must recover your vehicles when they cannot be fixed in situ and take them back to a vehicle workshop behind the lines—in much the same manner as a roadside breakdown service.

Finally, you must provide medical services to evacuate and treat your wounded, military police to police the areas that you bring under control, and tactical intelligence specialists who will question enemy prisoners of war and analyse your enemy's actions and intentions. You will also need civil affairs units to interact with (or provide) civilian administration in the areas you control, and mortuary teams to identify the bodies of, and evacuate, your dead.

These 'supporting' services are as important as your combat troops. Without medical services, your army's morale will plummet; without military police you will find your logistics under attack in your rear areas; and without tactical intelligence specialists you will only develop a weak understanding of what the enemy is doing.

As you saw in the chapter on training, all these capabilities must be brought together in combined arms formations known as battlegroups, brigades, and divisions (remember the rough rule of thumb is that there are three companies in a battlegroup, three battlegroups in a brigade, and three brigades in a division). These formations are underpinned by robust and secretive communications.

In a modern army, all communications will be encrypted, with more important (generally higher level) information receiving a higher level of encryption. In a modern army, you will not only be transmitting and receiving voice conversations—for example, giving orders over a radio to a company—but also sharing large quantities of data and telemetry between different command posts, intelligence and surveillance assets and sensors, and weapons systems.

This is precisely why you will need to develop a system akin to an encrypted, wireless internet between all of your units that is carried both across radio and satellite networks. Please heed the warning I gave you in Chapter 1: if you are fighting an enemy that can break into your encrypted networks, you must keep changing your cryptography codes, frequencies, and other methods of communication. This cannot be stressed enough: do not get lazy with your communications security, otherwise your enemies will know more about your armed forces than you do.

Creating an army

As leader, you will have endless supplicants and salespeople advising you on which capabilities you should include in your army. Some will advise you that you need more artillery, others more tanks, or perhaps that unmanned drones or increasing the size of your special forces will allow you to get rid of your reconnaissance troops. You will also receive much advice on how to structure your army, especially about the ratio between reserve forces and full-time forces, and whether you should sub-contract parts of your military to private military companies (mercenaries), or militias that you raise locally where you are operating.

My recommendation as you consider these conflicting opinions is always to look at the types of war that you will be fighting, and the enemies you will face. If your likely adversary has a land army of hundreds of thousands, with thousands of tanks, and massed artillery, then you will need some way of defeating this threat. If they have unsophisticated air defences, then build up your air force. If their air defences are sufficient, then you may invest in long-range stand-off missile systems. Or—and the reality is that military force is not cheap—you may need your own army of hundreds of thousands of tanks and massed artillery. The entire Cold War was an exercise in NATO and the Warsaw Pact observing each other's capabilities and developing counters. It will be no different for you.

Again, you must weigh up this advice in light of a realistic appraisal of the types of war you will have to fight (rather than wish to fight), and the likely enemies you will face. The reality is that reserve forces take time to become ready for deployment, and they always look larger and more capable on paper than what you end up with in your theatre of operations. If you choose to put key enablers into the reserve—say, logistics units—then you are hampering your ability to deploy your full-time army as well.

The problem with mercenaries and militias is that they are much harder to control than your own troops and harder to integrate across your entire force (they tend to do one job in one area).

The Myanmar Army uses militias extensively in the various wars it is embroiled in with rebel groups across its territory. The hired militias are indisciplined, commit war crimes, cannot be deployed beyond their home areas and spend much of their time enriching themselves. This is one of the reasons that the many conflicts that plague Myanmar have been going on since 1948.

Building an army takes years due to the time it takes to design and manufacture equipment, develop doctrine and train your forces. And as this process unfolds you must seek to maintain the balance of your forces: each capability has strengths and weaknesses, and so you must maintain appropriate ratios of each to the other for your force to work, all underpinned by sufficient logistics and communications. And because it takes such a long time to evolve your force, and militaries reflect the underlying culture of the countries from which they draw their troops, armies tend to be known for their different approaches and capabilities—their 'ways of war'. In the final part of the chapter, we will look at three differing ways of war—those of the West, Russia, and China.

Western armies generally have been known for having highly technologically advanced armies where to some degree technology compensates for the size of the force. This reflects the fact that European (and later American) societies have been at the forefront of the technology curve since the Industrial Revolution; it is now also symptomatic of those societies' non-acceptance of their own casualties. In the twentieth century and into the modern day, this is best exemplified by American dominance in highly-technologically advanced airpower, which can do huge damage whilst exposing the minimum number of your own people to any risk (unmanned aircraft or drones take this one step further).

Symptomatic also of the more democratic nature of their societies, Western armies tend to operate on a more decentralised command philosophy called Mission Command (outlined in Chapter 4), in which subordinate commanders are given tasks to achieve, but are not told how to achieve them, leaving them to seize the initiative. This command philosophy, in turn, requires more highly educated and trained troops—which the West also tends to have.

Coupled together these two aspects of the Western way of war—high level of technology and decentralised command—create the ability to fight what is called manoeuvre warfare. Manoeuvre warfare is about avoiding your enemy's strengths and destroying important—but often weakly defended points—like command and control centres. It is achieved by moving your forces around the battlefield rapidly in order to concentrate force when you need it and dissipate it at other times. Rather than attempting to destroy your enemy's force, you should seek to dislocate it, confuse it, and terrify it—thus destroying their motivation and morale to fight. In a sense, it is the opposite of attritional warfare.

Russian armies were historically characterised by their huge mass of poorly trained conscript soldiers, backed up by large amounts of artillery. They still are. As a result, Russian armies usually try to fight a form of attritional warfare, where they seek to grind down the enemy and destroy as many of their troops as possible, often with little regard for their own casualties, or civilian casualties.

As with the West, these aspects of the Russian way of war reflect their society. Russia is a large, poor country that is difficult to rule. Therefore a conscript army—usually drawn from the poorest peripheries of the country—solves several problems for the rulers in that it creates a large military to strengthen the central state. And if these conscripts get killed in war, then that—cynically—means that there are fewer young men in the poorer, distant rural areas primed to rebel.

The emphasis on artillery goes back to when Russia was becoming a great power, when it acquired a technological edge in cannons, so the use of artillery is bound up in the Russian military psyche with success, strength and greatness (much like technology is in the Western way of war). This psychological relationship with artillery even took on legendary status: when the Swedes defeated the Russians in 1700 and captured all of their artillery pieces, Tsar Peter the Great ordered every church in the country to give over some of their bells to be re-smelted into cannons.

The recent Russian invasion of Ukraine has followed a standard pattern for Russian wars, where huge amounts of artillery are used to pound enemy positions and destroy cities (often with civilians still living in them). These barrages are then followed up with poorly-trained massed infantry attempting to advance—infantry that usually takes very heavy casualties. These were similar tactics to those utilised in the Second World War, where Russia had the most casualties out of any country—up to 20 million people by some accounts.

Finally, the Chinese army (the People's Liberation Army or PLA) has another distinctive way of war which has grown out of its history as the revolutionary army that overthrew the nationalist Chinese government in the 1940s. As you might expect from such a populous country that until recently was very poor, the Chinese way of war used to privilege manpower over technology. And as befitting a revolutionary army, the Chinese often sought to advance columns of infantry into the enemy rear areas. These would then come together to create overwhelming local numerical dominance over enemy units, which would then be completely destroyed.

But the rise of China's power on the world stage during the twenty-first century has meant that China's military is evolving:

China now seeks to have a military that can fight short, decisive, high-tech conflicts much like the West has always sought to do. This is partly in reflection of Chinese understanding that its main competitor is the United States, who exemplify this technology-heavy way of war (and poorly trained massed infantry never do well against high technology militaries). But it is also a reflection of the fact that Chinese society is changing, becoming richer, and with much higher levels of penetration of technology.

Finally, and in a little appreciated second-order effect of the one-child policy, there are many families in China that only have one son. And because of the dates of the one child policy, many of these are now at military age. This fact makes China currently very leery of high casualty levels because of the destabilising effect on society if many families were to lose their only child.

As a result of these shifts, China is investing in high-end technology like hypersonic missiles, artificial intelligence, aircraft carriers and swarming drones. But as you will know by now, while technology is mandatory when it comes to effectively using military force to change your enemy's mind, on its own it is not enough. And because China has not been involved in any large-scale conflicts since the Sino-Vietnamese War in 1979, it is not clear whether China can transition its force to emphasising technology rather than manpower. To put it bluntly, it is fine having an aircraft carrier, but it is another thing entirely knowing how to use it. Whether China manages this evolution in its use of military power is one of the key questions in world geopolitics over the next thirty years.

The lessons that you should draw from this brief discussion of different ways of war are twofold. First, it is an extremely complex task to change how your army is structured and how it uses military force. Second, your army's way of war will significantly affect the types of war that it is able to fight. If—like the Russian army—your force relies on massed conscripted infantry, then do not try to fight short decisive campaigns. Your troops will not be highly enough trained or have the initiative to achieve your goals. Similarly, if you are a society that abhors taking casualties, then you must avoid wars that have a prospect of becoming long, drawn out and bloody.

This has been a brief overview of the structure and capabilities of your land forces: the decisive part of your military force. The next chapters look at those elements of military force that support your land army. First sea power, air power, and space power; and then cyber and information warfare.

1. The Strategy Stool

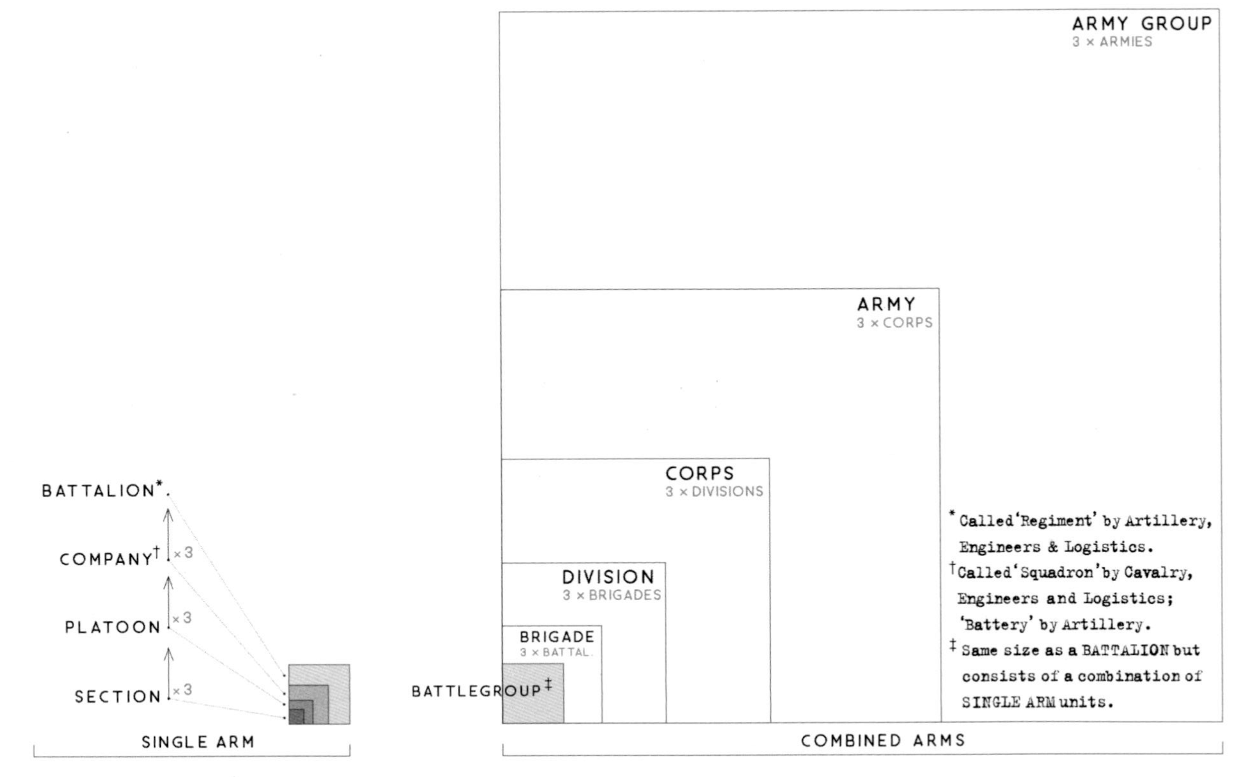
BATTALION*
COMPANY†
PLATOON
SECTION
×3
×3
×3
SINGLE ARM
ARMY GROUP
3 × ARMIES
ARMY
3 × CORPS
CORPS
3 × DIVISIONS
DIVISION
3 × BRIGADES
BRIGADE
3 × BATTAL.
BATTLEGROUP‡
COMBINED ARMS
* Called 'Regiment' by Artillery, Engineers & Logistics.
† Called 'Squadron' by Cavalry, Engineers and Logistics; 'Battery' by Artillery.
‡ Same size as a BATTALION but consists of a combination of SINGLE ARM units.
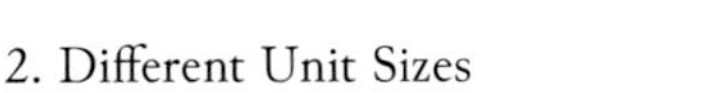

2. Different Unit Sizes

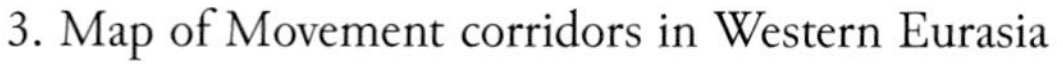
3. Map of Movement corridors in Western Eurasia

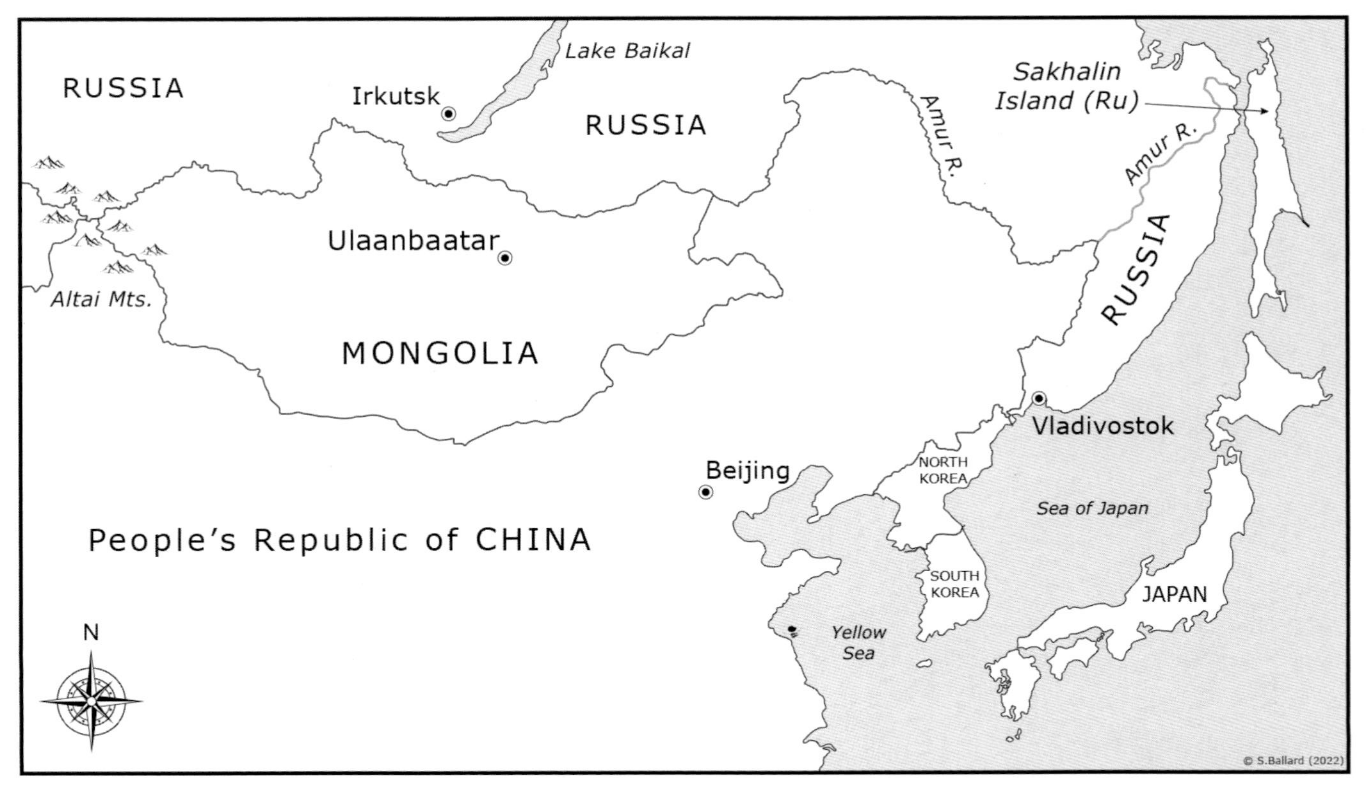

4. Map of the Sino-Russian border

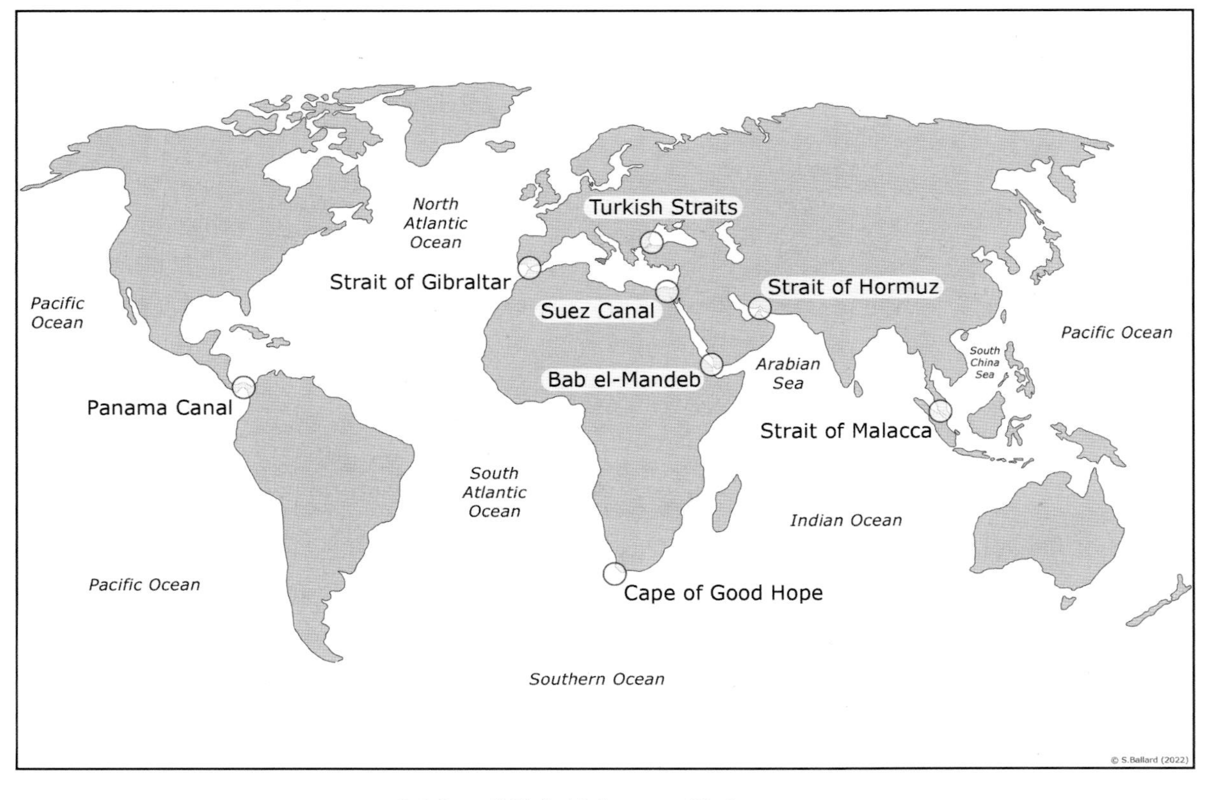

5. Map of Global Maritime Chokepoints

6. Map of Ukrainian Deception Operations, 2022

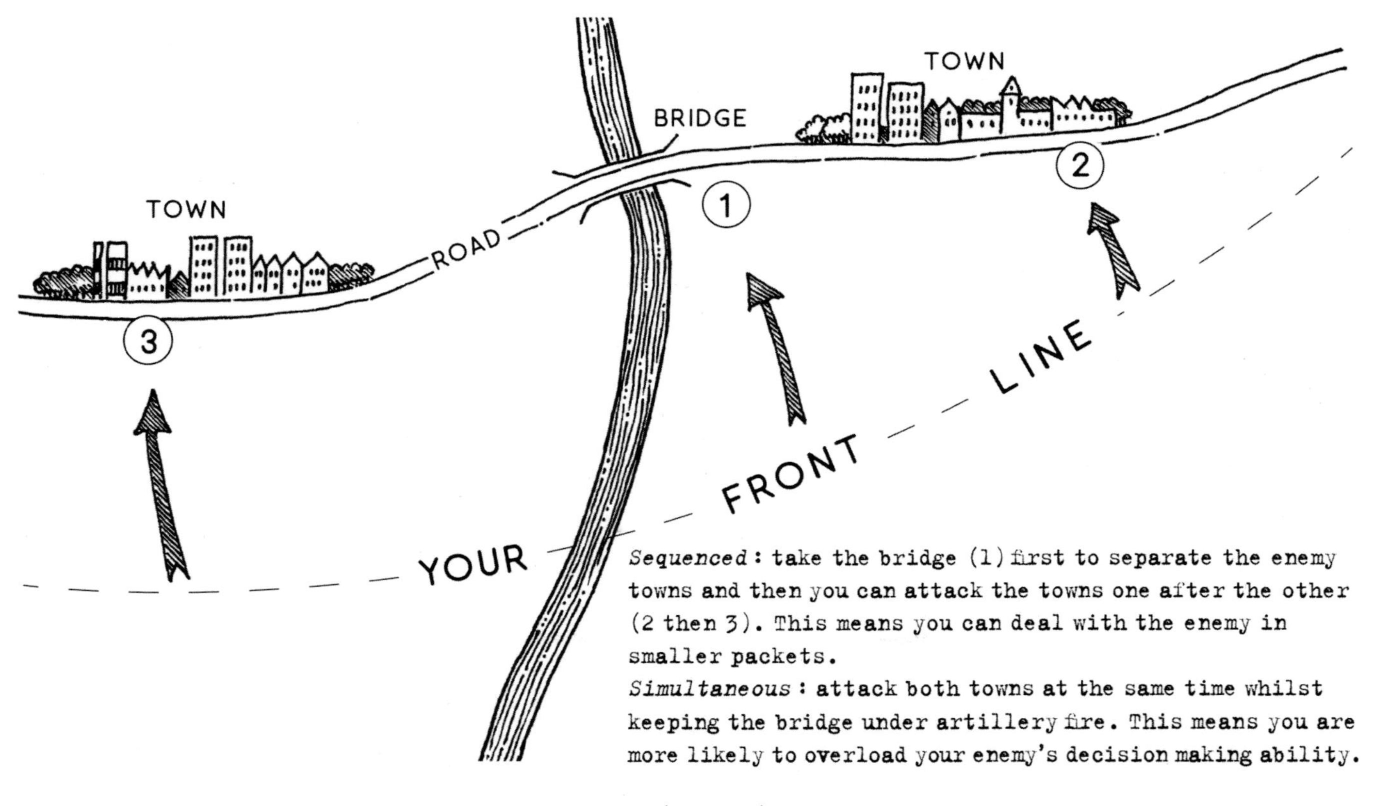

TOWN
BRIDGE
TOWN
ROAD
1
2
3
YOUR FRONT LINE
Sequenced : take the bridge (1) first to separate the enemy towns and then you can attack the towns one after the other (2 then 3). This means you can deal with the enemy in smaller packets.
Simultaneous : attack both towns at the same time whilst keeping the bridge under artillery fire. This means you are more likely to overload your enemy's decision making ability.

7. Sequenced vs Simultaneous

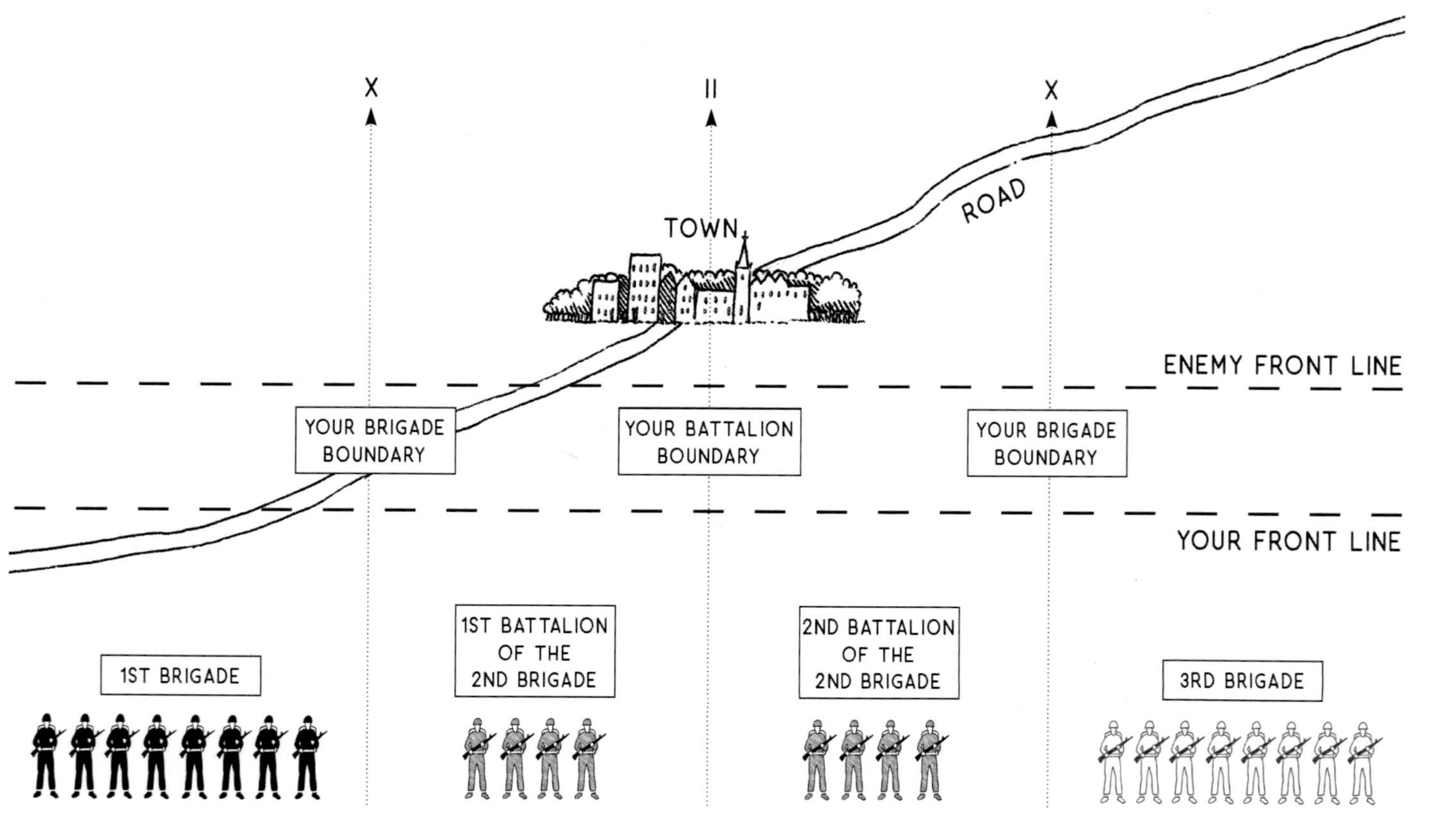

8. Battlespace Management

1

LEADERSHIP & COMMUNICATIONS

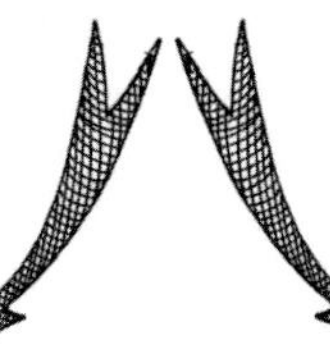

2

INFRASTRUCTURE

sea ports / roads / bridges

rail network / fuel lines

2

RARE ASSETS

radar / mine field clearance

air assets / engineering / bridging / intelligence

3

COMBAT POWER

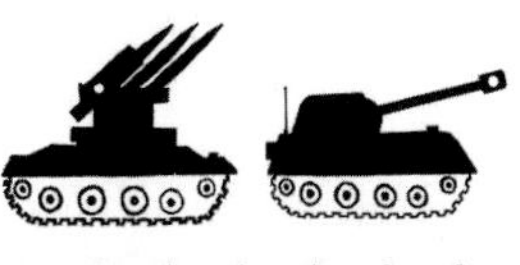

artillery / tanks / infantry

9. Targeting Hierarchy

In combat you are trying to make decisions faster than your enemy is. Anything you can do to slow their OODA loop down (e.g. by destroying their communications) and speeding yours up (e.g. by making faster decisions) will give you an advantage over them.

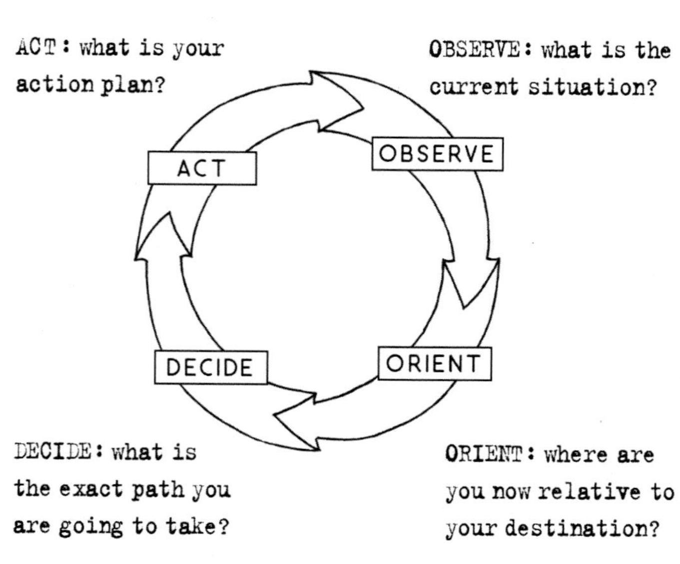
ACT: what is your action plan?
OBSERVE: what is the current situation?
ACT
OBSERVE
DECIDE
ORIENT
DECIDE: what is the exact path you are going to take?
ORIENT: where are you now relative to your destination?

10. The 'OODA' Loop

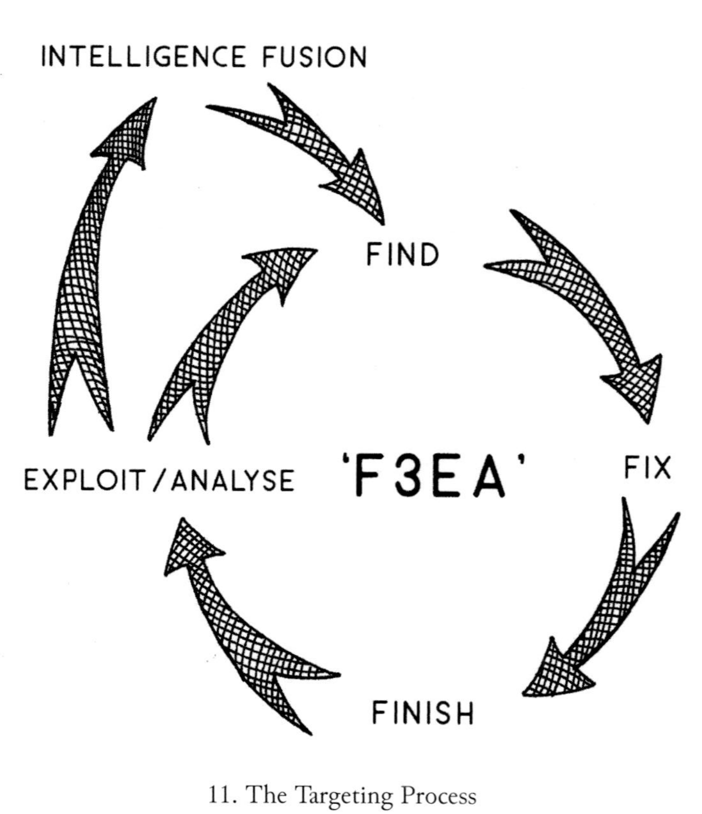
INTELLIGENCE FUSION
FIND
EXPLOIT / ANALYSE
'F3EA'
FIX
FINISH

11. The Targeting Process

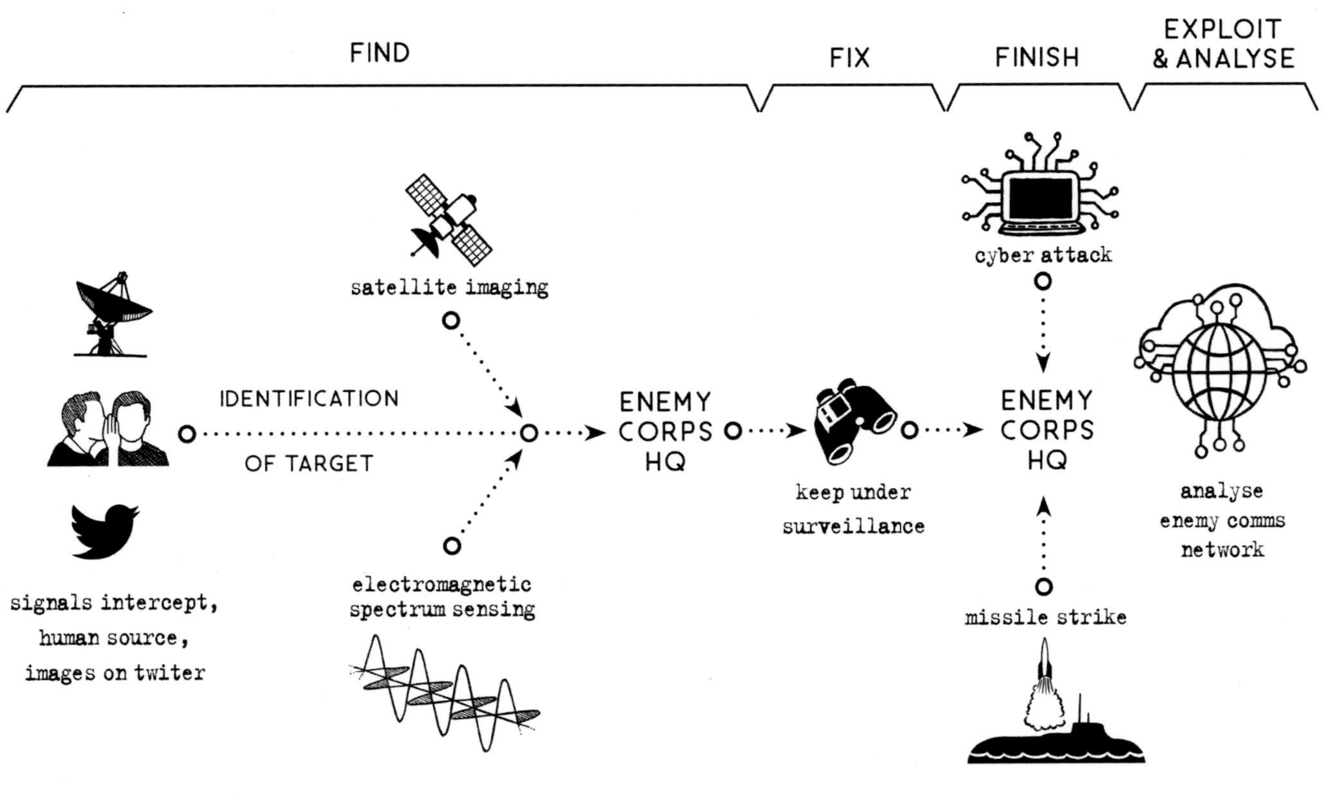
FIND
FIX
FINISH
EXPLOIT
& ANALYSE
satellite imaging
IDENTIFICATION
OF TARGET
ENEMY
CORPS
HQ
cyber attack
ENEMY
CORPS
HQ
keep under
surveillance
analyse
enemy comms
network
signals intercept,
human source,
images on twiter
electromagnetic
spectrum sensing
missile strike

12. Targeting Enemy HQ

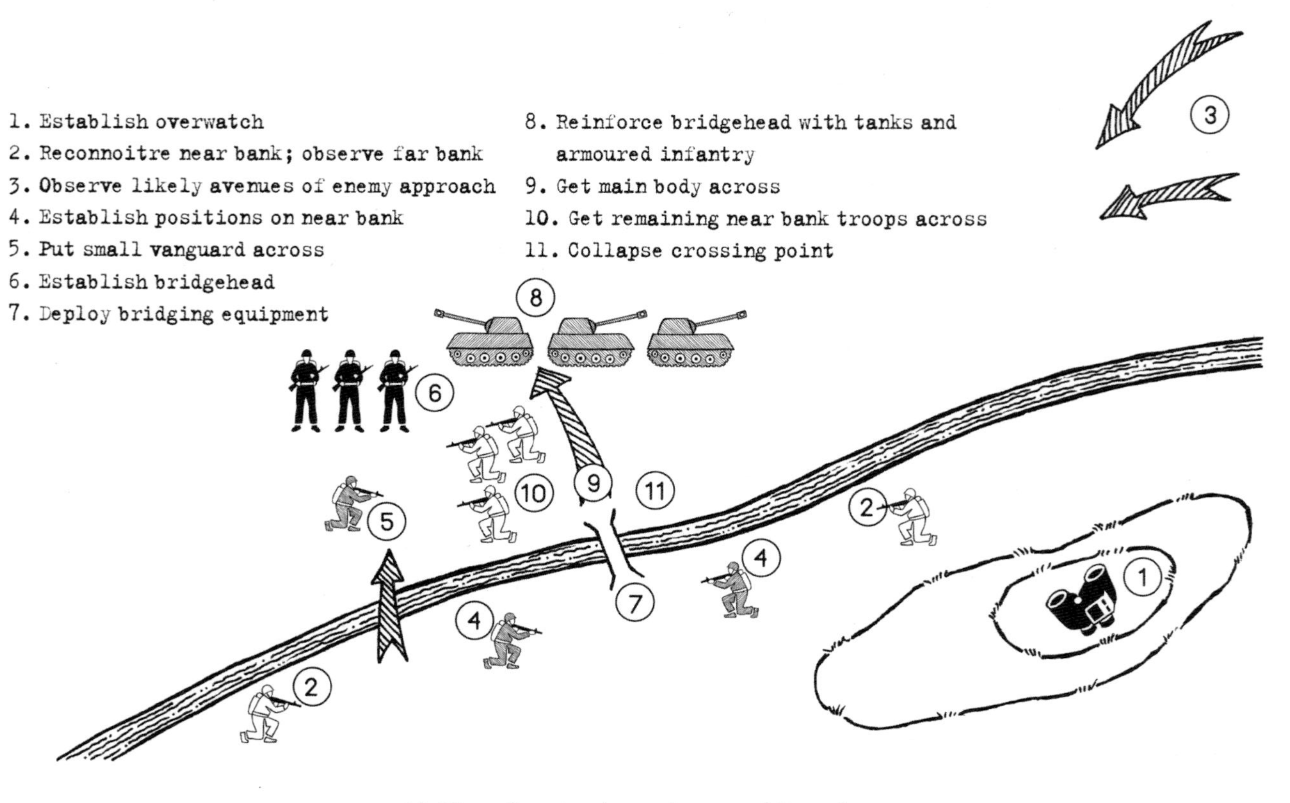
1. Establish overwatch
2. Reconnoitre near bank; observe far bank
3. Observe likely avenues of enemy approach
4. Establish positions on near bank
5. Put small vanguard across
6. Establish bridgehead
7. Deploy bridging equipment
8. Reinforce bridgehead with tanks and armoured infantry
9. Get main body across
10. Get remaining near bank troops across
11. Collapse crossing point
1
2
2
3
4
4
5
6
7
8
9
10
11

13. River Crossing by an Armoured Brigade

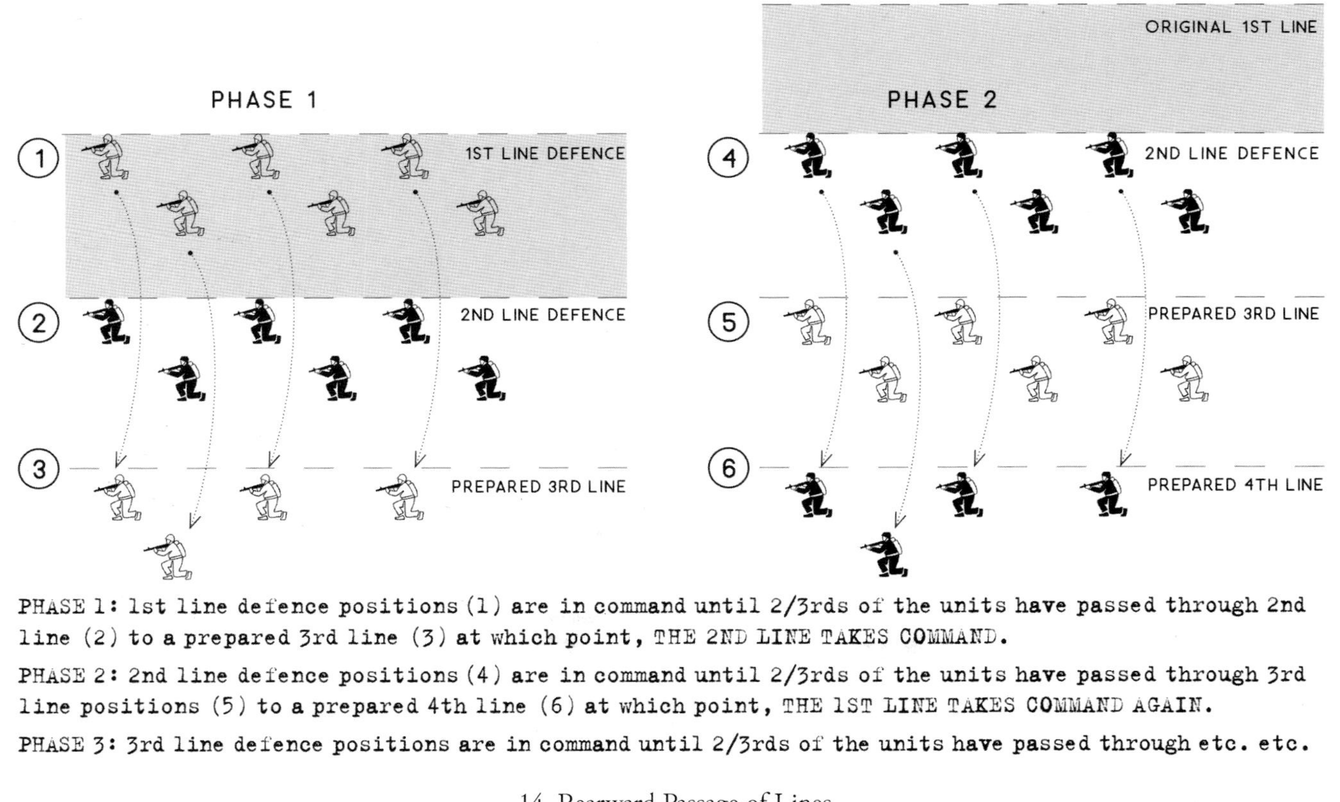
PHASE 1
1
1ST LINE DEFENCE
2
2ND LINE DEFENCE
3
PREPARED 3RD LINE
ORIGINAL 1ST LINE
PHASE 2
4
2ND LINE DEFENCE
5
PREPARED 3RD LINE
6
PREPARED 4TH LINE
PHASE 1: 1st line defence positions (1) are in command until 2/3rds of the units have passed through 2nd line (2) to a prepared 3rd line (3) at which point, THE 2ND LINE TAKES COMMAND.
PHASE 2: 2nd line defence positions (4) are in command until 2/3rds of the units have passed through 3rd line positions (5) to a prepared 4th line (6) at which point, THE 1ST LINE TAKES COMMAND AGAIN.
PHASE 3: 3rd line defence positions are in command until 2/3rds of the units have passed through etc. etc.

14. Rearward Passage of Lines

Each of your positions should cover at least one, normally two or three, of your other positions so they are mutually supporting.

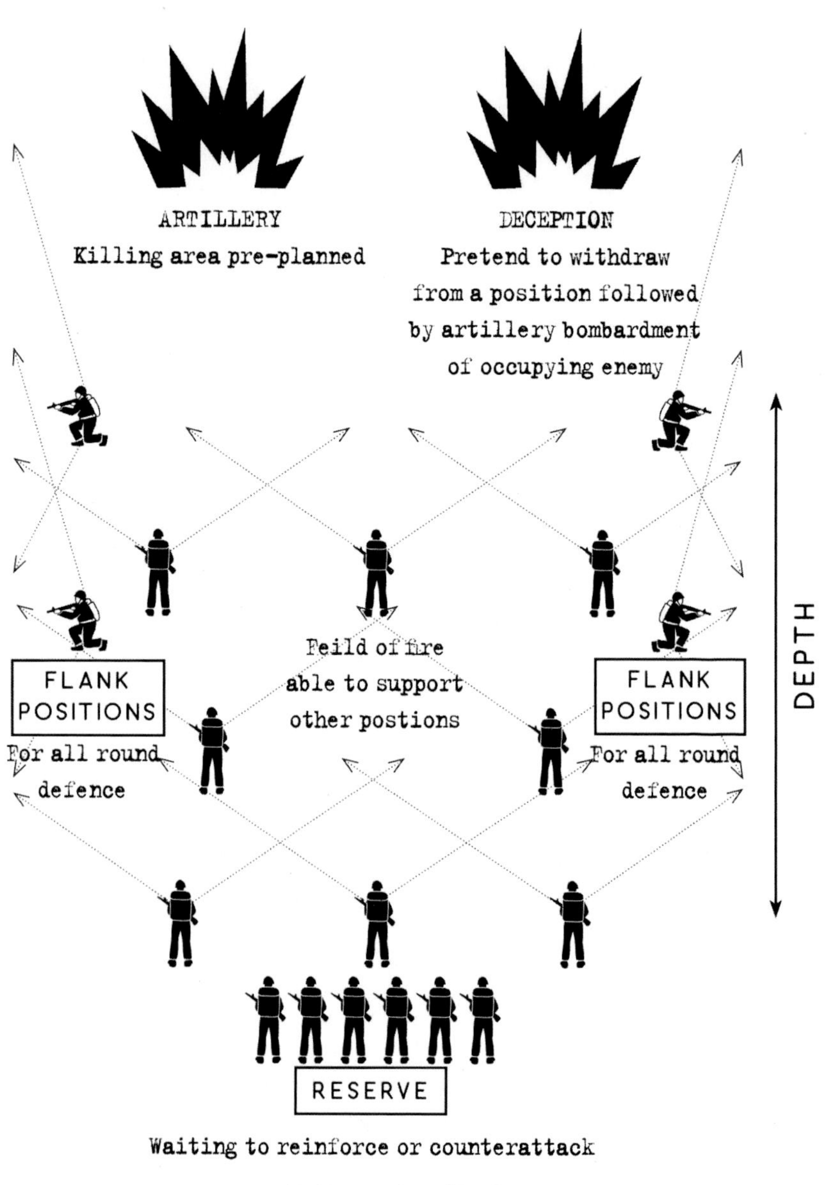

15. Principles of Defence

Arrange your forces in a dip where they are hidden from the enemy's view by a small rise. Locate an observation post to your rear on higher ground where they can observe the enemy's approach. When the enemy crests the small rise to your front, destroy them.

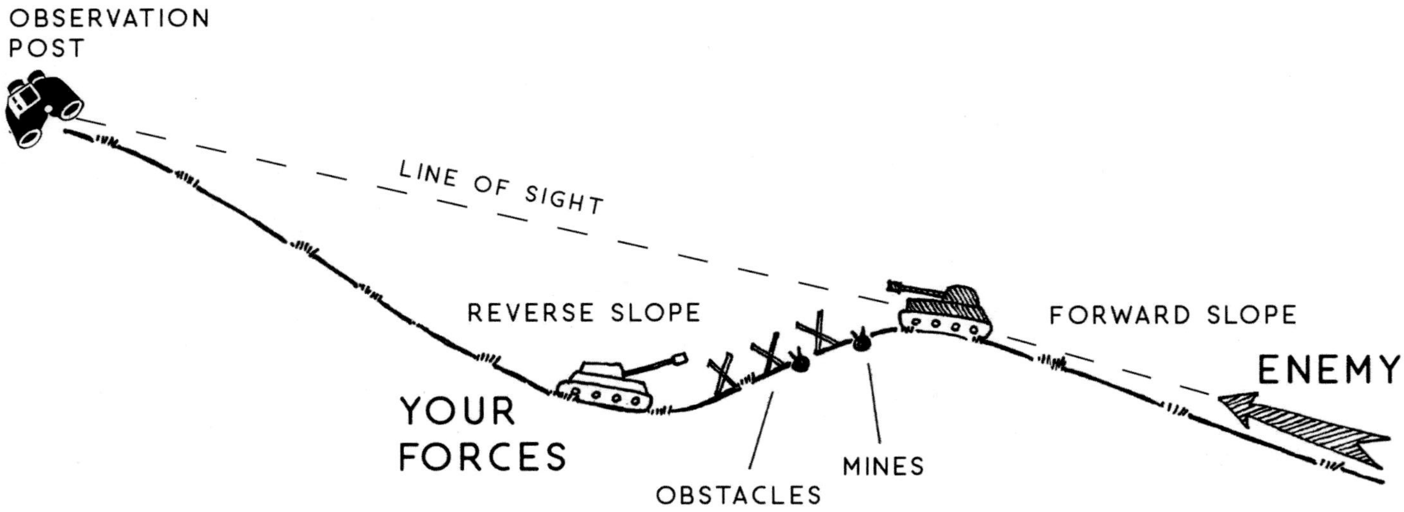

16. Reverse Slope Defence

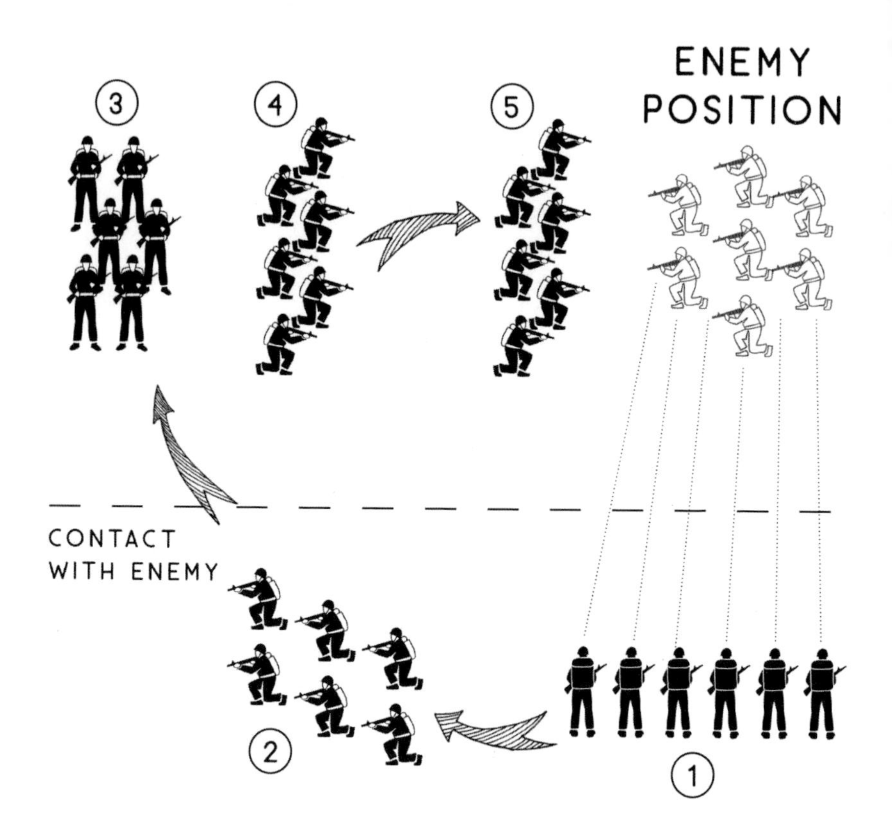
ENEMY POSITION
3
4
5
CONTACT WITH ENEMY
2
1
APPROACH
1. Establish baseline to suppress enemy
2. Establish flanking force
3. Place reserve
4. Assaulting force
5. Assaulting force gets near enemy position, supressing force keeps its fire in front of assaulting troops

17. How to Conduct a Basic Assault

6.

SEA, AIR AND SPACE

Since antiquity, military power on land has been supported by naval forces. In the last 100 years, land armies have been supported by air forces. And in the last fifty years, all three domains—land, air and sea, have been supported by military activities beyond the earth's atmosphere. In this chapter, we will explore these three aspects of military power.

Until now, we have been focussing on manpower—its training, morale, and operational organisation—as a key determinant of military power. Equipment has been somewhat subsidiary, if nonetheless important, to the overwhelming importance of the human element. In the land environment, you are seeking to equip the man or woman. But in the sea and air domains (and even more so in space), you are seeking men and women to run the equipment.

The overriding reason for this is the vast expense of naval and air force equipment: the price tags for single items of equipment can run into billions of dollars. Developing air and sea power involves designing and producing equipment at the very forefront of what is technologically possible, thus few states can afford to maintain militarily successful forces at sea, in the air, and particularly, in space.

A word of warning. While it is true that at sea, and in the air, we tend to focus on technology over manpower, you must remember that war is still about humans, and about human psychology. As commander of your forces, it will be natural for you to see your naval and air forces in terms of equipment—how many ships you have lost, versus how many enemy planes you have shot down. But as with all warfare, we are seeking to change the enemy's mind. Warfare is not a spreadsheet; it is a conversation.

And so the question you should be asking is not how many enemy ships or planes will you destroy. Instead, you should determine how you wish to affect the enemy's psychology, and which military activity you should be conducting in the maritime, air and space domains to achieve the desired impact (naturally, this will often include the destruction of enemy ships and planes, but not always). The difference is subtle, but important, and each of these three domains of warfare generate different types of support to the ultimate political goals of defeating your enemy. This idea of focussing on enemy psychology rather than technology remains true even in the case of cyber warfare.

In this chapter, we will first look at how you can best deploy military power in the maritime, air and space domains to support your land operations, and to change your enemy's mind.

Maritime power

Humans have been building up maritime military power since the struggle for control of the Mediterranean between 800 BCE and 150 BCE. Before then there had been some riverine 'navies', but this was really the first occasion on which developed navies began contesting maritime space. This period culminated in the Romans gaining dominance of the sea after the Punic Wars. Two themes emerged from this era that have continued to shape how military power is used at sea.

The first theme is that the dominant global power at any one time controls the known world's sea and ocean space. The Romans achieved mastery of the Mediterranean Sea and other nearby seas like those around the Britannic archipelago. Over a thousand years after the close of Roman power, the Portuguese, followed by the Spanish, the Dutch, and the British, established a series of quasi-global or global empires all of which relied upon dominion over the oceans. Since the twentieth century, this role has been played by the United States.

That the dominant power controls the oceans is obvious in one practical sense—it enables you to move your soldiers and military supplies around the globe to where they are needed. In this very fundamental way, maritime dominance facilitates your military control of the land. But this leads us to the second theme: control of the seas and oceans is always used to facilitate the passage of free trade. The great global empires have at their heart been trading empires, which has required naval dominance to move goods and people.

Trade and navies go hand in glove: the accumulation or generation of wealth through trade enables the projection of maritime power, and that very same maritime power enables trade to flow by keeping shipping routes free from attack by hostile powers or pirates. Beyond the territorial limit of 12 nautical miles from the coast the laws of any one state do not apply, and the 'high seas' are only weakly governed by international law.

The formal lack of control over the oceans—probably the greatest commons available to humanity—necessitates policing to enable ships to pass freely (so-called freedom of navigation). This role has always been fulfilled by the pre-eminent military power of the time primarily because it is in their interests to do so, and this creates a positive benefit for everyone else. This is highly significant: 80 per cent of global trade is carried on the oceans.

One of the easiest ways to analyse control of the world's sea space is to look at who controls, or provides hegemony over, the global maritime chokepoints (see Fig. 5). These are eight points on the global shipping map that are essential for ensuring the free passage of world trade. Some of these are critical for a single commodity: the Strait of Hormuz between the United Arab Emirates and Iran carries 30 per cent of world oil, and the Bosphorus 25 per cent of world wheat exports (as well as 3 per cent of its oil). Others are critical for linking regions: the Malacca Strait carries the 25 per cent of world trade that flows between eastern and western Eurasia (Europe & the Middle East to China), and the Panama Canal links east and west coast America, and the Atlantic and the Pacific Oceans.

Who controls the global maritime space is a vital factor as you consider your plans for war. Do you rely on any of these maritime chokepoints, and if so, who controls them currently? Is this great power—and it will inevitably be a great power—allied to you, or at least likely to allow you to continue to transit the chokepoint with your trading vessels (to allow your economy to continue to function) and your military vessels (to supply your troops)? If not, think back to the chapter on logistics—can you survive sanctions, and a blockade on your country's trade?

Next, as you consider building your maritime forces, you should decide at which scale you wish to project maritime power. A brown-water navy, the cheapest option, aims to protect your rivers and lakes with small gunboats and patrol boats. However, brown-water navies are little more than police forces for your internal waterways, or where a river forms the border with another state, a customs force. Laos, Paraguay and the Central African Republic have small brown-water navies with a few patrol boats equipped with machine guns.

A green-water navy aims to control and police the coastal zone of your state, and in some cases can project power into the seas and oceans of the neighbouring region. Most coastal states have some sort of green-water navy that enables them to protect their coastline, stop illegal fishing and resource extraction, and project power out to the 200-nautical-mile Exclusive Economic Zone that every coastal state is granted under international law. The key determinant of a green-water navy (as opposed to the truly global blue-water navies) is that they are tied to their home country which they rely on for supplies, and potentially air cover.

Navies at the top end of the green-water navy category often have amphibious ships able to deploy troops from sea to land, helicopter (and in a few cases, aircraft) carriers, and a mixture of destroyers and frigates to carry out their missions. Current examples of green-water navies—and countries can gain or lose naval power as their economies wax or wane—include Brazil, Australia, and India. Australia, for instance, has helicopter carriers and landing ships, alongside submarines and frigates, that enable Canberra to project power into the Pacific Ocean.

The largest navies, capable of self-sustained global operations at short notice, and for indefinite periods of time, are known as blue-water navies. Currently, the only undisputed blue-water navy is the United States Navy. With almost 500 ships, comprising 11 aircraft carriers, 78 submarines, 72 destroyers, 22 cruisers, and a host of other capabilities including launching large-scale amphibious operations, it can launch and sustain indefinitely maritime operations anywhere on the planet. Traditionally the three other blue-water navies were the British Royal Navy (73 ships) and the French Navy (c. 100 ships), both of which are now too small to conduct sustained global operations, and the Russian Navy, which is so riven by corruption, mismanagement and poor maintenance as to now be largely ineffective.

Although it has grown in recent years to become much larger in terms of tonnage, the Chinese Navy has not yet demonstrated the ability to conduct military operations globally. In addition, many of the ships are new and it takes a long time to train your personnel and develop an understanding of how to use your new capabilities in wartime. In short, the position of the Chinese Navy is analogous to that of the Chinese Army—it has recently gone through a period of rapid growth, has a lot of new equipment, but is yet to be tested in war. Fighting wars is like anything else: you tend to get better with practice. Militaries that go to war often, tend to be good at it, if only for the amount of time they spend doing it.

If you are planning—and can afford to build—a blue-water navy, then you must consider the balance of capabilities that you require. This depends on the likely tasks you want your navy to carry out, and the enemy that you expect to face. In the latter half of the twentieth century, many large navies opted to have large aircraft carriers from which they could fly fast jets. Aircraft carriers enable you to project military force anywhere on the planet because the planes can attack both enemy ships (this is very rare and was last done when the US Navy sank an Iranian ship during the Iran-Iraq war in the 1980s), and enemy targets on land (very common, and it probably happened this month somewhere in the world).

Aircraft carriers are extraordinarily expensive. Their costs not only include the carrier itself (c. US$13bn for the largest), and the planes to go on them (c. US$100m each), but due to their vulnerability each carrier requires an escort group of ships to protect it from enemy attack. Taken together this 'aircraft-carrier group' (which will likely also include a submarine) requires multiple ships with the collective ability to conduct anti-submarine warfare, air defence (including anti-missile defences), and anti-ship warfare.

Many navies have found to their cost that the prestige and capability that comes with an aircraft carrier ends up gutting the rest of your navy because all other ships end up being subsumed into escort duty for the carriers. There are also very open questions now as to whether aircraft carriers (and other large ships) can successfully defend themselves against more advanced modern missiles that fly at multiples of the speed of sound ('hypersonics'), swarms of micro-drones, or flotillas of micro-ships or micro-subs (discussed below). This is an ongoing debate in the British MOD, which recently commissioned two large aircraft carriers just at the moment when drones and hypersonic missiles have come into their own.

Beyond aircraft carriers, naval surface ships have traditionally come in two broad categories that have had changing names over the years. The first category is that of a smaller, faster, more lightly armed ship used for interdiction and patrolling sea lanes and keeping them open—these days they are likely to be called a frigate or a destroyer. Second, navies have a category of bigger, slower, more heavily armed ship used for taking on enemy ships and bombarding the shore—these days likely to be called a cruiser. Most surface ships are also highly capable air defence platforms that can offer air defence over hundreds of square miles including the land.

In the twenty-first century, with improvements in propulsion technology, and the replacement of naval guns with missiles, the distinction between the two types of naval ship has become blurred and many navies just have one category of surface combatant. When you are deciding what type of surface combatants you want, look to the capabilities you will need. Do you want to conduct shore bombardment? Attack enemy ships? Defend against enemy aircraft? Or attack enemy submarines? This will help you build up your surface fleet.

The larger navies can conduct amphibious operations—that is, deploy land forces from a ship straight onto a littoral (usually a beach). Opposed amphibious operations are probably the most complex types of military operation possible due to the difficulties in getting enough troops ashore fast enough and then continuing to supply them. While this is all going on you must coordinate a complicated air, land and sea battle in a concentrated piece of terrain, while also striking targets deep in the interior of enemy-held territory to stop them reinforcing and pushing you back into the sea. It is extremely difficult to do well.

If you still decide that you want the ability to conduct amphibious operations then you will need to procure and train on amphibious landing ships (which deposit the troops on the beach), amphibious supply ships (which keep them supplied), and possibly helicopter carriers (which is another way of getting your troops from sea to land quickly, and defending those already there). You will also need to consider how to defend your amphibious operation from the air, so you will need an aircraft carrier, or ships with air defence capabilities.

If you can make this work—and not many navies can—you will have a fearsome psychological weapon that you can use to overawe potential enemies, or at the very least cause them to tie up many of their land formations defending their coastline from the threat of your amphibious attack. Currently, only the United States—in its Marine Corps—has the breadth and depth of the capabilities required to conduct even a medium-sized amphibious operation that has to contend with serious opposition.

The Chinese navy has built up a large amphibious force—presumably aimed at retaking Taiwan at some point. But could it conduct an amphibious operation across 100 miles of sea that is opposed by the Taiwanese and their allies the United States? I would pay careful attention to US (and now Australian and British) attack submarines, and to the submarine fleet that the Taiwanese are currently building up—this will enable the island state and its allies to sink the Chinese amphibious fleet well before it gets to Taiwan.

Submarines were originally developed for sinking enemy ships unseen, but this role has currently declined. Only two ships have been sunk by submarines since the Second World War: an

Argentinian cruiser by the British in the 1982 Falklands War, and an Indian ship sunk by the Pakistanis during the 1971 Indo-Pakistani War.

Nowadays, it is the threat of submarines, and the psychological fear that they engender among those in surface ships, that carries out this role: submarines are regularly used to screen an area of ocean ahead of a task force or aircraft carrier battlegroup, in much the same way that infantry is used to screen ahead of a tank, thus denying that area of the sea to the enemy. This fear is increased if an enemy has nuclear-powered submarines. These use a small nuclear reactor to propel the submarine, as well as power its systems (including making water and air), meaning that they can stay submerged for several months at a time, needing only to surface for food supplies or maintenance.

In modern navies, however, submarines are most likely to be used as platforms that can sit off the enemy coast unobserved and then launch missiles (or occasionally special forces teams). They also act as highly specialist reconnaissance and surveillance platforms that surveil enemy coastlines and follow and listen to enemy military assets like ships and aircraft. Several countries are developing unmanned submersibles which are particularly suited for reconnaissance gathering roles.

Submarines are also involved in protecting and/or penetrating the huge undersea optic fibre cables that form the backbone of the internet, and hence carry 95 per cent of the world's communications. Be highly aware of communications security—which undersea fibre optic cables does your data pass through, and do any of your enemies have access to those cables? The recent spate of attacks on oil pipelines and undersea fibre optics in the North Sea and around Scotland—most likely perpetrated by Russia—demonstrates the importance of this threat.

Finally, submarines are considered the platform most likely to survive a nuclear war, and so several countries use them to carry their nuclear deterrents. We will discuss this further in Chapter 8.

Navies take a long time to build up, and so the decisions you make now about what kind of navy you want will affect your country for the next fifty years. If you aspire to a blue-water navy, then this means generating the full spectrum of capabilities described above, including an aircraft carrier (or carriers) and an amphibious capability. You will need a fleet of logistics ships to supply your navy flotilla (as many ships again, in reality). And finally, your navy must have sufficient scale that it can continue to operate globally even when one in three of your ships is being maintained or refitted (considered a normal maintenance/deployment schedule). Global, blue-water navies are not for the faint-hearted!

If your country cannot afford this outlay and is content to settle for a green-water navy, then you will have many more (cheaper) options. If you only want to attack land targets in support of your army, then a fleet of surface combatants armed with precision missiles will suffice. If your aim is to reconnoitre secretly and to deny areas of the sea to enemy shipping, then submarines are a good bet. If protecting your coastline is a priority, a large fleet of small, fast gunboats might be appropriate. If you require the capacity to interdict your enemy's military supply vessels, then submarines, surface combatants, or aircraft might be appropriate. Bear in mind that the costs will start to spiral when your enemy has a competent surface fleet combined with submarines, or an effective naval air capability.

In the near future, many navies will opt for larger numbers of smaller vessels. This takes advantage of advances in computing technology (it wasn't possible previously to develop such small capable ships), and they offer a way to avoid the potential vulnerability of very large, expensive ships like aircraft carriers. The natural progression of this technological trend towards more, smaller platforms, is for countries to develop swarms of fleets of small, unmanned surface or subsurface vehicles. As ever in the military world, this will force competing militaries to develop capabilities that can counter this threat.

The capabilities that you acquire should always be rooted in your likely enemies, and the concomitant threat that you face, and particularly the psychological effects you wish to have on them. Will sinking their ships terrify the remainder of their navy? Will blockading their ports cause food shortages and riots? Do you want to demonstrate that you can dominate the sea, and restrict their shipping? Will you use missiles to destroy enemy communications networks? The options are endless, but because of the lag time in building ships, you must think now about what you intend to use your navy for many decades hence.

Air power

Air power has one key advantage: if you have air dominance, and your enemy doesn't, their land army cannot operate effectively because you will be able to target and destroy the critical parts of their ground forces.

If you have control of the skies—and by this I mean not only that there are no enemy aircraft, but also that their anti-aircraft defences have been destroyed—it will not take you very long to destroy your enemy's logistics system, or its communications systems, or its command posts. As you know, this will be the end of their ability to operate as an army. It is for these reasons that most militaries seek to achieve air dominance in the early stages of a war, because it then allows them the freedom to reconnoitre, strike enemy targets, and move around the warzone as they wish.

The downside of air power is that it costs an unbelievable amount of money, and as a result aircraft and helicopters are always in short supply, unless you are among the very richest of nations. Training a Chinook pilot takes at least three years. A B2 stealth bomber costs $2.1bn, and $135,000 per hour to run: rarity and expense will dictate how you should use air power. Great care is taken not to risk aircraft or pilots, and aircraft, aircrew, and aviation infrastructure like airfields and radars are among the highest priority targets. And as we have already stated, air power is limited in that it cannot on its own win a war—you will always need land forces for what is called the 'decisive' effect.

Air power is predominantly about reconnaissance, that is finding out where the enemy is and what they are doing, and delivering ordinance, that is dropping bombs and missiles on your enemies. In the last twenty-five years, several militaries have also developed unmanned aircraft ('drones') that conduct these two tasks. Aircraft and helicopters also have a secondary role of transporting light things, including personnel, to and around the battlefield.

Only very rarely is air power about engaging enemy aircraft or helicopters in aerial combat dogfights. This has happened a few times since the Second World War, mostly in the Korean and Falklands Wars. Over the last thirty years, it has hardly happened at all, largely because the aircraft of very advanced nations use stealth technology to avoid enemy radar, and combinations of other technologies to engage enemy targets from a long way away. Nations that lack such technologies tend to find their planes are destroyed on the ground if they go to war with an advanced nation. The very few air-to-air kills that have happened over recent decades have been where low-technology nations have fought wars with other low-technology nations.

Traditionally, aircraft have had a role in the collection of all the different types of intelligence outlined in Chapter 1 (except for, obviously, human intelligence). Reconnaissance aircraft were able to take photos, collect signals, electronic and emissions intelligence. Loitering undisturbed over a battlefield enables a huge amount of information to be hoovered up about enemy troop numbers, equipment, and their intentions. In the early twenty-first century, much of these roles have been given to satellites and unmanned aerial vehicles, also known as drones.

The most important versions of these planes are called Airborne Early Warning & Control (AEW&C) which can not only use their sensors to look for enemy ground units, but also track enemy aircraft, missiles and artillery. Usually they also serve as a secure communications relay for your ground forces, and as a command-and-control platform that keeps track of all the enemy targets across a wide area, and help you prioritise which ones to attack first (they are manned by a crew of up to ten people to carry out the various different roles). These are a very high value target for your enemy and so you must be careful to protect them well.

In contemporary warfare this collection role is called Intelligence, Surveillance, Target Acquisition and Reconnaissance, or ISTAR. And on a modern battlefield, you should aim to have several ISTAR assets overhead with a variety of different sensors, including the ubiquitous video link to your headquarters that enables you to view the battle in real time and make decisions based on what you see. Not every ISTAR platform has all the sensors, and so layering them on top of each other generates a more detailed picture of the battlefield for your generals to exercise command and control. These days, most operations rooms of small units (companies of around 100 troops) will have ISTAR downlinks to screens to enable commanders to follow the battle.

Because of their smaller size, payload and range, helicopters can provide specialist reconnaissance roles such as aforementioned, but over a much smaller area. The most specialist example of this is those helicopters which conduct anti-submarine warfare, often launched from aircraft or helicopter carriers. They are equipped with a suite of sensors to detect submerged enemy submarines—radar, sonar, infrared and magnetic anomaly detectors—and sometimes with the weapons required—depth charges—to either destroy the submarines or to force them to the surface where they can be more easily attacked.

The second main role of air power is striking enemy targets. This broadly splits into three categories—strategic bombing, missile attack, and close air support to your troops on the ground. Strategic bombers—like the American B-52, which has been in service since the late 1950s and has been used in every war that the US has fought since then—can fly many thousands of miles (the B-52 has a range of 8,800 miles) and can drop up to 30 tonnes of bombs.

This type of high-altitude bomber should be used in the initial stages of a conflict to attack enemy command and control, and logistics—once enemy radar and air defences have been destroyed by either missiles or aircraft that avoid enemy radar detection. If you want to destroy large targets like supply dumps, or enemy airfields, then large strategic bombers are your best option.

These types of air operations are highly complex and involve hundreds of people from intelligence staff assessing the targets, through ground crew preparing the aircraft, to operations staff who plan and deconflict the airspace, keep an eye on enemy aircraft or defences, and help navigate the bombers to target. If the latter is a long way away, then a tanker aircraft may need to be arranged, or intermediary airfields or crews put at the ready. In short, getting bombs on targets is a system of systems, and only a small handful of countries is able to do this successfully at large scale.

Alternatively, many countries have guided missiles that enable them to accurately attack enemy targets with, say, a 500kg warhead. These missiles—generically called cruise missiles—are used to attack pinpoint targets like command bunkers, radar sites, ships and bridges. They can be deployed from aircraft, submarine, ship, or land-based launchers and the type of missile you decide to develop or buy depend upon what sort of enemy you envision being at war with (that is, your launchers must be able to get within range of the target). Compared to other ways of delivering munitions, they are relatively cheap, costing $1–2m each.

Missiles come in a huge variety. This includes different warhead sizes, different ranges (usually up to about 800 miles, but with some that can travel several thousand miles), different speeds of flight (ranging from subsonic, through supersonic, to hypersonic, and different flight modes (do they hug the ground to avoid radar or fly high in the atmosphere to escape ground-based anti-missile systems?). If you can afford it, you should procure several variants—with different ranges, payloads, speeds and flight modes—because this will complicate matters for your potential enemies who will have to generate an anti-missile capability to counter all of your missile types.

Finally, you must consider close air support for your army. This is where your ground troops can call upon aircraft or helicopters to deliver cannon fire, missiles or bombs directly onto the enemy. This is usually during battle and so requires a high degree of precision and exact coordination between your ground and air forces. For this to work, you will need specialist troops on the ground (called forward air controllers) working side by side with your infantry and tanks to direct your aircraft and helicopters onto enemy targets and to stop them from hitting your own troops.

Some of these varieties of close air support can be very effective. Helicopters are very useful for destroying enemy tanks. They can hide behind hills whilst being fed targeting data by a camouflaged and hidden air controller, before popping up and destroying the enemy tank (or tanks in the case of the Apache helicopter—they can engage sixteen targets at the same time), then moving back behind the hill and into cover. It is because of the existence of such effective weapons systems that you must strive to gain air dominance early into a war, or if you are unable to dominate the air space, you must try to keep it contested—that is, under no overall control, thus thwarting your enemy's ability to attack your ground forces at will.

Increasingly, the main functions of air power—reconnaissance and striking enemy targets—are being taken over by unmanned aerial vehicles (UAVs). Drones can be both fixed-wing, like aircraft, which allows for high-altitude, long-endurance flight, or rotary-wing, like helicopters, which make shorter flights, closer to the ground, and are able to hover. They have huge advantages over manned aircraft: they are much cheaper (approximately $30m for a high-altitude, fixed-wing type like a 'Reaper'), and you can deploy them freed from the fear of losing your pilots if they are shot down.

There is an argument that drones are too easy to use, thus making countries more likely to use military force in circumstances where they otherwise wouldn't or couldn't. This is almost certainly the case in the recent 'Global War on Terror' where the United States overused drones to assassinate those who it saw as terrorists all over the world. Because they failed to solve any of the underlying reasons driving the conflicts, those terrorists were merely replaced by others—usually even more embittered—thereby necessitating further drone strikes.

Large fixed-wing drones can be used in both a strategic role—hitting individual high-value targets like leaders, or communications nodes—as well as providing close air support to your troops on the ground. In both of these roles, it is a great advantage that they can loiter for hours above the battlefield (most aircraft don't have the fuel for this, and in any case want to get on and off the target quickly to avoid danger to the aircraft or pilot). These large drones also have a suite of reconnaissance sensors, meaning that one drone can reconnoitre the battlefield—or maritime environment—and then follow up and strike its own targets.

Drones have become progressively smaller over time, and military micro 'quadcopter' drones look exactly like the ones you can buy from online retailers like Amazon. In fact, many rebel groups are buying drones from online retailers and either using them straight out of the box for reconnaissance, or slightly modifying them with explosives to turn them into effective strike assets (either by dropping the explosives like a bomb, or—because they are so cheap—acting as 'suicide' drones). US forces in Syria, for instance, of which there are about 900, are regularly attacked with cheap modified commercial drones, sometimes in swarms.

This technology is changing the tactics of wars at the time of writing. In the 2020 Nagorno-Karabakh war between Azerbaijan and Armenia, small Azeri drones were used to destroy first the Armenian air defences, and then all, or most of, the Armenian armoured vehicles and tanks. Armenia was effectively defeated by the Azeri drone fleet, from which they had no ability to defend, and was forced to accept a ceasefire with the loss of much territory. If you are likely to face such an enemy, I would recommend that you start thinking now about anti-drone defences that are able to shoot down multiple micro drones simultaneously.

Similarly, in the Russo-Ukraine war that began in 2022, Ukrainian forces have used small military drones to great effect, destroying Russian supply vehicles. In addition the Ukrainians have used micro commercial drones to reconnoitre the battlefield and find Russian units to attack, as well as conducting specialist missions like observing where Ukrainian artillery is falling so that the fire can be adjusted and made more accurate—traditionally this would have needed an observer team within a kilometre or two of the enemy. It proved devastating for the poorly-trained Russian forces and forced them to think again about basic military skills like camouflage and dispersal of vehicles.

Looking to the future, you can expect major military powers to develop and deploy thousands of tiny micro drones at the same time. These will be networked together and controlled through an artificial intelligence algorithm that is distributed across processors held on all the drones. This weapon would be able to flock and swarm, much like starlings. Moreover each drone would be its own suicide bomb (rather than launching a munition), and in the event of the destruction of any single drone the network would continue to function. Weapons like this will be very difficult to counter, let alone defeat, and if you find that your enemy has successfully developed something similar, it would be best not to go to war until you have worked out how to counter it.

Finally, control of the air enables you to conduct airborne operations—that is, dropping large numbers of troops (and their equipment) out of aircraft with parachutes, or delivering smaller numbers by helicopter. These are almost as militarily difficult as amphibious operations, and for the same reasons—it is very difficult to achieve a high enough troop density on the target, and to subsequently supply forces that you have delivered by air.

The most recent example of their attempted use was during the Russo-Ukraine war where, in the very early stages, Russia attempted to conduct airborne attacks on Hostomel airfield near the Ukrainian capital Kyiv. Although the helicopters bringing the initial assault force managed to land, the Ukrainian ground forces mopped up the disgorged troops. Because of examples like this, very few countries maintain parachute forces, let alone use them. On the other hand, as a psychological weapon, their deployment, or rather fear of their use, can be very effective.

As you can see, air power is vital in enabling a successful land operation, and only a fool would embark on a land war without control of the air.

Space power

If you are from a very rich nation, then you must also think about the military uses of space. Over the last fifty years, the largest powers have developed satellites that provide reconnaissance, communications, and navigation capabilities to air, sea and ground forces. There are approximately 320 military satellites in orbit, with over half deployed by the United States.

Operating in space offers several advantages to the powers that are able to do it. The very best reconnaissance satellites can resolve an image at 5 or 10cm, which is more than enough to distinguish one weapons system or piece of equipment from another, or to count the number of people in a place and work out what they are up to. You will be unable to read a newspaper headline, as some have claimed—for that you would need to put a drone overhead with a high resolution camera.

Some reconnaissance satellites can also monitor stray radio communications and sense electronic and emissions to assess, for example, the capability of an enemy radar system. Others are specifically designed to detect and provide early warning of the trajectory of intercontinental ballistic missiles (more on this in Chapter 8). The great advantage of reconnaissance satellites though is that they are much less likely to be shot down than aircraft, or intercepted, and if you have enough satellites in complementary orbits you can generate continuous coverage of the area or target that you want to reconnoitre.

Communications satellites will enable you to communicate securely with your military wherever you have satellite cover (they are particularly secure because they are line-of-sight, meaning that in order to intercept them you must get between the military unit on the ground and the satellite—i.e. have some sort of plane overhead). At present, only the US, China, Britain and France have global, secure military satellite communications.

Russia used to operate at that level, but it lags technologically behind the Western powers and China. It has also been increasingly unable to replace satellites as they have neared the end of their working lives (in part due the sanctions that it has faced since its 2014 annexation of Crimea). The limitations in its secure communications capability severely hampered its 2022 invasion of Ukraine enabling Western powers to develop an understanding of Russian capabilities and intentions which they passed on to Kiev, allowing highly successful targeting of Russian command and control nodes and personnel.

The final main use of satellites is highly accurate navigation, down to 2–5m. This allows your personnel to navigate but also helps multiple precision weapons systems to find their targets. At present, there are four global networks of satellites that provide navigation services. They are owned by the United States, the European Union, China and Russia (again Russia is having difficulty maintaining its fleet, and its coverage is degrading).

If you are unable to launch your own constellation of navigation satellites, you must make sure that you are allied to one of those powers, and that your relationship is sufficiently robust to survive any negative consequences of any war that you have planned. It is, quite simply, impossible to imagine launching a war in the modern age without precision navigation, especially if your enemy has access to the capability.

It is for this reason too that China—and others—have investigated and practised how to 'spoof' GPS signals in certain areas, so that the navigation systems on (for example) ships show a different location to that actually occupied by the ship, or 'jam' them to completely cut out the navigation signal. This technique was potentially implicated in a spate of collisions involving US Navy ships over the last decade.

In response, the US Navy has again started teaching its navigators how to navigate using pens, paper, a sextant, the stars and the sun. Every space power is concerned about how much they rely on satellites for (particularly) communication and navigation, and how vulnerable they might be to anti-satellite missiles, lasers, or potentially other satellites, especially constellations of micro satellites designed to bring down entire satellite networks.

It is also worth highlighting that military activities in space require ground stations, abilities to launch assets, highly skilled staff and secure data links. You can take out your opponent's space operations (and they yours) without destroying their space assets!

The military uses of space beyond the earth's atmosphere are a key area of great power competition in the twenty-first century. What shape that competition will take is not yet clear. Although all the major space powers have signed the 1967 Outer Space Treaty that places some limitations on the military use of space—namely a proscription on weapons of mass destruction, and the non-military use of the moon and other celestial bodies—military activities per se are not restricted in orbit.

This leaves open the door for the eventual placement of conventional munitions in orbit, which will enable one country to bombard another from space, with very little warning. Even currently, several countries are developing hypersonic missiles which use the edge of space to fly at many multiples of the speed of sound (up to Mach 17–13,000mph—in some cases). These can either be tactical hypersonic missiles which can evade most current defences on, for example, ships. But they are also being developed at intercontinental range, where the missile is boosted up to the edge of space orbit, and then glides down onto its target at hypersonic speeds giving a flight time between Moscow and Washington of approximately fifteen minutes. Russia, for example, has tested several of these weapons.

As we look to the future—across all three domains of sea, air and space—there are certain trends that are emerging. Whereas previously, countries sought to have large, prestige systems—big ships like aircraft carriers, expensive jet aircraft like stealth bombers, and large constellations of satellites—advances in technology are making them very vulnerable to enemy attack. You should consider whether a greater number of cheaper, smaller and unmanned systems offers a more survivable and effective military capability—particularly as micro ships, drones and satellites that are networked and controlled through so-called swarming artificial intelligence are developed over the coming decade.

7.

INFORMATION AND CYBER

Much has been written about how modern wars will be fought with information rather than tanks, and by cyber-attacking enemy computer systems rather than deploying infantry. No doubt: the twentieth and twenty-first centuries have seen huge advances in communications technologies which create significant advantages, but also huge vulnerabilities, in our military systems as well as our societies.

But despite all the hype, these technologies have had variable effects on the conduct of war. 'Information warfare', after all, is what has been called propaganda since the beginning of time, albeit now there are now vastly more effective communications systems through which to disseminate your messages.

As the 2022 war in Ukraine demonstrated, this heralding of a new style of warfare has been overly optimistic and exemplifies one of the fallacies of war outlined in the introduction: getting bewitched by a new technology that allows us to wish away the normal rules of warfare.

In the case of cyber and information warfare, the normal rule that is being abjured is that the land environment is where wars are won or lost. This fallacy is made yet more attractive by the relative costs of 'hard' military power, and operations in cyberspace: 'virtual' operations can be a hundred or a thousand times cheaper than deploying armoured divisions or launching air strikes. These types of attacks also have the key advantage of being sometimes deniable, the further to sow confusion among your enemies.

That said, cyber and information warfare are like military power in the sea, air and space domains: very useful or even essential in supporting the decisive effort on land, but not able to win wars on their own. When done well, cyber and information warfare can have significant effects on enemy psychology, which as you know, is the main target in warfare. Ultimately, however, directing lethal violence against your enemy is far more persuasive then bombarding them with information: tanks will always beat tweets.

This chapter will explain to you how best to use cyber and information warfare to support your strategic and tactical aims and operations. A brief caveat: I will discuss only on the use of information and cyberspace operations above the threshold of warfare and combat. I will not comment or offer any advice on the ubiquitous use of these activities below the threshold of war—for example the regular claims of election interference by one country into another's democratic processes, or cyber attacks that take government systems offline, or shut down websites and internet services—mostly because that would be another book in itself.

Strategic information operations

There are four main audiences that you are trying to communicate with in strategic information operations: the 'world' audience, your own home population, and the enemy's population and leadership.

Communicating with the first two audiences is about maintaining consent for your actions and is of the utmost strategic significance. Usually both audiences will receive the same narrative, because projecting different narratives to these two audiences will discredit you in the eyes of both (there are a few exceptions to this—like North Korea—where the domestic communications landscape is so restricted as to enable the North Korean government to say different things to different people).

Communicating with the third audience—the enemy population—is done primarily to drive a wedge between them and their leadership, thus attempting to weaken or shatter the bonds—outlined in Chapter 2—between government, population and military. And lastly, communicating with the enemy leadership—both political and military—is very different from the first three audiences, and is much more likely to involve deception of some kind and/or the coincidental use of force or the threat of force to reinforce your message.

At the highest level, your information operations are a distillation of your strategy. That is, the ability to describe your strategy as a narrative and for it to be considered reasonable to the widest global audience goes some way to helping you judge if your strategy is realistic and achievable. In many senses, strategy and communication are the same thing. After all, violence is simply another way of communicating, or a way of emphasising the credibility of what you are communicating. The chances of your strategy succeeding, your international logistics system working, and your home morale being maintained, are all critically dependent on framing your war in a way that is considered fair and reasonable by a neutral observer.

As you will recall, one of the most important things that you must do when deciding your overall strategy is to fix upon the narrative that explains your actions.

The world population, and the other leaders of countries that they influence, must broadly see your war as a just war, otherwise you risk losing international support at a time when you need it the most to ensure the supplies of critical military technologies and commodities required to sustain your war effort (see Chapter 2: Logistics). Your population must see it as a fair and reasonable war because they will suffer the deprivations of conflict, and must regard these setbacks as worthwhile in light of the war's narrative (see Chapter 3: Morale).

The narrative likely to gain the most widespread worldwide acceptance is that of defending your homeland against an unprovoked aggressor seeking to annex your country to theirs. Perhaps the hardest is to be that unprovoked aggressor—in the twenty-first century, other populations and states tend not to like unilateral redrawing of the world's map. Other wars—say to topple a government to install a different one more sympathetic to your country—fall somewhere in between and whether others consider them fair and reasonable usually depends upon factors beyond your control.

As a key part of your strategy process, work hard to frame your conflict in a way that is acceptable to outsiders. This takes intellectual and emotional effort, because you have to see your actions from others' perspectives—but is inexpensive when compared to the costs of getting it wrong. It will pay dividends in terms of ensuring the overall consent for your operations, your international logistics and supply, and your home population's (and hence your military's) morale.

The most effective way of generating information operation narratives that seem fair and reasonable to neutral outsiders is—exactly as with strategy formation—to form an organisation with diverse perspectives, with red teams whose job it is to challenge the status quo, and with the odd maverick here and there. Groupthink is as much an enemy to good information operations as it is to strategy—indeed they are the same thing—and it is through guarding strongly against groupthink that poor narratives, overly centred on your own perspective and unreasonable to others, are weeded out.

There are several advantages to framing your strategic narrative very early on in your strategy formation and planning process, not least in that it may force you to alter your overall strategy, should you realise that your narrative is unacceptable to third parties.

The third target population that you might hope to influence—that of your enemy's home population—is very difficult to influence, as most people naturally tend to believe their own leaders over and above information emanating from other countries, particularly those that they are at war with. That said, most states will endeavour to communicate with the enemy's home population. Usually, the narrative will be a derivative of their wider-world, audience-facing narrative, with particular emphasis on points that may drive a wedge between the enemy leadership and their population. You should make efforts to communicate with your enemy's population, but not to expect too much success.

A good example of shaping a narrative to maximum strategic benefit was that of the United States-led coalition during the First Gulf War in 1990. At the beginning of August 1990, Iraq invaded and occupied their smaller neighbour to the south, Kuwait. George H.W. Bush, the US President at the time, very skilfully pulled together a coalition of thirty-five countries, including numerous Muslim countries, using the narrative that the coalition's aims were simply to evict Iraqi forces from Kuwait, and not to attack nor to occupy Iraq.

The narrative was well accepted enough to receive the backing of a UN Security Council resolution, keep the coalition together even in the face of Iraqi missile attacks on Israel (which might have brought Israel into the war, thus fracturing the coalition as many Muslim countries would not have wanted to have been on the same side as Israel), and received enthusiastic domestic support in the United States (5 million people lined the streets to welcome the troops home after the war).

The narrative was important because it reflected a strategic reality—made all too clear by the disastrous US-led invasion of Iraq in 2003—that changing the political leadership of Iraq was not a problem that could be 'solved' by military means without huge, unanticipated and negative consequences both for the Iraqi population, and for US prestige and strategic interests both in the region and around the world.

By comparison, the long American wars in Vietnam and Afghanistan had grandiose narratives attached to them (defeat of communism, defeat of terrorism) that did not match the reality, namely that American troops were fighting guerrillas who were resisting them as occupiers. Russia made the same mistake in Ukraine in 2022 by arguing that it was fighting 'Nazis'.

The final audience that you should influence with your strategic information operations is the enemy's leadership. The big difference is that for the first three targets—world opinion, your domestic population, and the enemy's population—you must shape your information operations so that they reflect reality as a reasonable outsider would consider it to be. But when communicating with your enemy leadership, signalling intent and deception should often take centre stage.

A useful way to think of this is that your information operations should—when backed up by violence or the credible threat of it—affect the decision-making calculus of the enemy leadership. Essentially you are using narratives and force to manoeuvre psychologically around your enemy, so that they make decisions that are not ultimately in their interests or are in your interests.

There are two main modes of communicating with the enemy leadership in war: signalling your intent to deter them from doing something and deceiving them as to your true intentions so that you gain an advantage in the conflict. Both types reflect the basic psychology of strategy and war and should underline clearly to you the unity between strategic information operations and strategy.

If you find yourself having to deter an enemy from taking military action, you must carefully calibrate what you say to them (and the world), and how you back up your narrative with force or the threat of force (which is sometimes accompanied by a small demonstration of your ability to project force).

For this type of signalling to work you must be as credible as you possibly can, both in narrative and in demonstration of military intent. For example, if your intelligence agencies tell you that there is a threat of invasion from a neighbouring state, then you can only deter them successfully if your threats of turning it into a bloody and costly mistake at least broadly match the balance of power. If there is a complete mismatch, then your narrative and strategy of deterrence will fail.

A good example of where deterrence worked was during the Cuban Missile Crisis in 1962. The USSR had positioned missiles on Cuba, enabling it to strike the mainland United States without the US being able to detect the missiles before they struck. In response, the US instituted a naval blockade of the island, and stated that any further ships heading there would be sunk. They also demanded that all missiles on the islands be removed. Once the USSR realised that the US considered the placement of Soviet missiles on Cuba as a core security red line, and believed that it would use military action to maintain this position, it backed down (the US removed some missiles from Turkey in return).

You should compare this successful example of deterrence with the attempts by the United States and its NATO allies to deter Russia from invading Ukraine in 2022. In the days and weeks preceding the invasion, and as it became clear that Russia was about to attack, NATO warned Russia against invading, threatening economic sanctions and international isolation if it did so. At the same time, the NATO allies explicitly said that they would not intervene militarily to defend Ukraine (which you could consider to be the opposite of deterrence). This was interpreted as a green light to invade by President Putin of Russia.

The other type of strategic signalling that you should consider is convincing your enemy you are doing something or about to do something, when in fact you are going to do something else: in other words, deceiving them. It is the cheapest way to win a war. If your information operations increase the ambiguity in your enemy's mind about what you are planning to do, then they are working. And similarly to deterring, in order for your deception to be successful, your words must be matched up with physical actions that underline what you are saying (unless you are planning on conducting unattributed—or black—information operations, which we will discuss below).

An excellent example of this comes from the Russo-Ukraine war. Some months into the conflict the Russians announced that they would focus on taking a region in eastern Ukraine called the Donbas. They thinned out their forces in the remainder of the country and began a general offensive there. The Ukrainians encouraged the Russians to think that this was the most important area of military activity in Ukraine—President Zelensky announced that it would be the key battle to save the country, the Ukrainian forces began for the first time in the war to publish casualty statistics, and Ukraine regularly appealed to its Western backers for more forces whilst pointing at the scale of the battle for the Donbas (see Fig. 6).

Whilst everyone's eyes—and especially Russian eyes—were focussed on the Donbas, the Ukrainian forces quietly started an offensive in the south of the country aimed at taking back the much more strategic city of Kherson. They managed to reconquer far more territory, of a greater strategic value, in the south than they lost to the Russian forces in the East. And once the Russians realised what was happening, the Ukrainian government announced a 'million-man' offensive in the south to retake

Kherson, forcing the Russians to redeploy units to the south ... whereupon the Ukrainians started attacking in the south-east, the East, and the south, to hide from the Russians the axis of their eventual assault.

One of the most successful ways of deceiving your enemy is to conduct unattributed—or black—strategic information operations. This is where you deliver information to your enemy via a third party in a manner that obscures the fact that the information came from you. One of the most successful—or perhaps the most famous, because certainly there are many successful deception operations that are still secret—occurred during the Second World War when British Intelligence were able to convince the Germans that the invasion of Sicily was a feint, designed to draw German forces away from Greece and Sardinia.

Ingeniously, they obtained the body of a dead homeless person, dressed him up as a Royal Marines officer, and placed documents on him emphasising Greece and Sardinia as targets of the invasion over Sicily. The body was then released from a submarine near the Spanish coast where it was picked up by a fisherman, with the documents eventually making their way to German intelligence. Signals intelligence confirmed that the Germans had fallen for the deception, and when the allied invasion of Sicily came, German reinforcements were sent to Sardinia and Greece.

However you plan to set the strategic information context for your war—the narratives for the world audience, your home audience, the enemy population and leaders—you must take into account the huge changes that have been wrought to the communications and information environments by the internet. (The internet now reaches 63 per cent of the world's population, up from 50 per cent in 2017, and some areas like Northern Europe have penetration rates of 98 per cent). An analysis of these changes would—and does—fill books and is part of the reason that information operations became so lauded as a new way of doing war in the first two decades of the twenty-first century.

However, these momentous changes to the communications landscape can be summarised in two words: disintermediation and fragmentation. Disintermediation is the removal of 'middle-men': previously people received their information about the world around them via a limited number of press and media outlets. After the widespread adoption of the internet there are thousands of sources to go to to receive information about the world around you. Importantly, many of these may be anonymous, not who they purport to be, or highly partisan. It is also possible—indeed common—to directly listen to someone's narrative from the other side of the world in a way that simply would not have been possible before the invention of the internet.

These processes have led to a splintering and fragmentation of the communications space—in short, to chaos. And such chaos offers significant advantages to your country if you want to disrupt the information operations of another. But this very fragmentation works against you if you want to project a unifying narrative that different audiences will believe and follow. In short, the internet offers advantages to challengers who wish to disrupt the status quo but makes it much harder for established powers to maintain coherent, widely believed narratives. We can see this clearly in the actions of Russia as they attempt to disrupt American hegemony over the second decade of the twenty-first century.

Russia, almost more than any other nation, invested heavily in information operations over this period. It rapidly expanded its overseas news broadcaster, Russia Today, and ramped up its activities online, particularly on social media, where they specialised in fake accounts whose job was to further fragment the communications landscape by injecting 'disinformation' or 'fake news' into the environment. The purpose of this further fragmentation was to create chaos and make it harder for Western governments to project coherent, believed narratives; indeed to govern. It is important to note that Russia did not create the fragmented environment, but merely exacerbated its fragmentation. And the Russian information activities didn't advance an alternative world view, but merely muddied the water to discredit Western narratives.

Into this chaotic information landscape, Russia has also dabbled with interfering in elections in the United States, the United Kingdom and France, among other countries—partly to favour certain candidates and/or policies, but mostly to seriously discredit the concept of democracy, which is obviously an anathema to autocratic systems. In many ways, Russian strategic information operations mirror the activities that the Russian state uses on its own population: create many competing narratives, so nobody knows what is true anymore, therefore creating the need amongst the population for a strong leader who brings clarity.

This Russian model of strategic information operations was very successful from 2014–17, but is becoming less and less successful as populations in the West—who were their target—have become more savvy about their use. Western governments have also learnt methods of countering this type of strategic disinformation operations—not least announcing Russian intentions before they carry them out, thus preparing the target (Western) populations that they are about to be the target of a disinformation operation.

As internet penetration and usage increases, the field of strategic information operations is constantly evolving. Most strikingly, the use of social media within warzones has completely changed the breadth and speed of reporting on conflicts, allowing combatants, observers and analysts to shape the strategic narratives of the war in real time. For instance, there is a trend of online expert influencers using social media to gather information about ongoing conflicts—from on-the-ground eyewitnesses, as well as other sources—and then to disseminate their own wider strategic analyses of the events in question which are then picked up by the traditional media.

These expert analyses are now starting to shape and lead the coverage of the traditional media who still rely on the old model of war correspondents in different locations, thus giving them excellent pin pricks of detailed information, but little overall view of the strategic direction of a conflict or wider geopolitical event. That is, social media is now shaping the overall narratives and perception of conflicts and geopolitical events. You should consider developing this capability if you wish to shape world opinion in real time.

Strategic cyber operations

Cyber operations encompass a range of highly secret activities that aim to shut down or damage your enemy's computer systems, both military and civilian. Because of the relatively low cost involved, a number of countries including all the UN Security Council members, plus Iran, North Korea, and others have developed varying degrees of offensive and defensive capability. Most of the time it is very hard to conclusively work out who carried out a cyber attack, and in any case most cyber attackers deny using such weapons. This 'deniability' does offer advantages in that you can damage your enemy's capability or send signals to them in ways that fall short of war, but it does increase the potential for misjudgement. As you might expect, there are innumerable ways of gaining access to and disrupting your enemy's computer systems: the main ones follow below.

Perhaps the most well-known type of cyber attack is that of the computer virus or malware. This is a (usually small) computer programme that is introduced into target computer systems either through a computer network or, more likely, physically introduced into the system on a USB drive (because critical systems are often air-gapped and not connected to wider computer networks like the internet).

The malware then has access to the computer system and can carry out its attack whether that is to freeze the system and lock its normal users out, delete or encrypt sensitive data, or hijack core computer functions like the processor so the system seizes up. Many of us have been subjected to these types of viruses on our home computers.

Sometimes viruses or malware are designed to exploit something called a zero-day. A zero-day is a previously unknown vulnerability in a piece of software—like the operating system or a word processor. Zero-days stem from the complexity of the software manufacturing process: it is inevitable that there are inadvertent vulnerabilities when even day-to-day software like Windows 10 has 50 million lines of code.

Examples of zero-day attacks include the Sony attack, where details of forthcoming movies and actors' personal details were released, and the RSA attack, on a security company, which allowed the attackers to gain remote control of the company's systems.

These vulnerabilities can be of any type: some allow access, some allow the software to be shut down, others allow users to be locked out—in short, the vulnerability can be in any aspect of the software's operation. This also means that if the vulnerability is in a very common piece of software—an operating system, for example—then the virus or malware can infect systems worldwide uncontrollably, causing indescribable, unplanned damage contrary to what might have been intended.

Once their products are on the market, software vendors are constantly attempting to discover and patch them. But vulnerabilities that have been discovered but are unknown to anyone (zero-days) are regularly traded on the (dark) internet fetching prices of up to US$10m (some software vendors also operate in this marketplace buying up the vulnerabilities in their own software so they can fix them).

The final main type of cyber attack is known as a distributed denial of service attack or DDOS. In a DDOS, many different computers around the world are harnessed (usually via some sort of computer virus or malware) to connect to a particular computer or website—say, the enemy government's website—to overload and shut it down. Estonia, for example, repeatedly suffers from DDOS attacks of Russian origin.

An important reality of these cyber attacks is that they are usually short lasting. That is, once you have managed to access your enemy systems and bring them down or otherwise disable them, it is usually possible for your computer experts to isolate the infected system, patch the vulnerability, and get your systems back online within a few hours, or at most a few days.

From this point forward, the vulnerability that was exploited—particularly if it was a zero-day—will be known to you and software manufacturers, and that style of attack cannot be used again.

These two characteristics of cyber attacks—that they are short lasting, and 'single-shot'—shapes how and why they are used. Broadly, you should consider two types of usage. The first is to knock out critical strategic-level computer systems at the start of a broader attack, so that your enemy struggles to respond to you effectively. Ideally, by the time they have their computer systems back up—say, those systems that control their air traffic, or a cluster of satellites—you will already have gained a decisive advantage with your land forces.

The second type of usage revolves around attacking your enemy's civilian infrastructure—maybe taking their television channels off air, or shutting down their power plants—and are designed to cause panic and fear among an enemy population, thus creating pressure on their leadership.

This second type of large-scale attack is yet to be carried out during a war, and nor is it clear how well it will work. For instance, other types of mass attacks on civilians like aerial bombing often result not in enemy civilians panicking and demanding peace as is often thought, but in a more resolute attitude determined to defeat the enemy that has bombed their cities. It is not predictable what dynamics massed cyber-attacks on civilian infrastructure might cause.

And because cyber weapons can travel across the internet and infect systems that were not the original target, they may have unintended consequences—say, shutting down hospital systems, causing people to die, and souring world opinion against you. In short, we have not yet experienced a war with large scale cyber-attacks, and so no-one is entirely sure what would happen.

At the very minimum, you must protect all your military computer systems, alongside your critical national infrastructure—a bewilderingly long list of water providers, power producers, air traffic control systems, internet and communications backbones, banks, supermarket distributions systems, and so on. Most countries that take cyber defence seriously have established a government agency that works with utility providers and private organisations that provide critical services to the population helping them to improve their cyber defences.

Probably the most effective strategic cyber attack was when the United States and Israel attacked Iranian nuclear facilities at its Natanz plant—although neither country has yet formally declared that they were responsible. Although not a use of cyber weapons in war, it is worth recounting here as it demonstrates how a critical enemy system can be disabled.

A piece of malware—later nicknamed 'Stuxnet'—was introduced to the systems at the plant most probably via a free USB stick that one of the Iranian nuclear scientists picked up at a trade fair. This malware exploited four separate zero-days software vulnerabilities to burrow through the Microsoft operating system and reach a piece of Siemens software that controlled the 5,000 centrifuges that were enriching uranium at the facility.

The Stuxnet virus gradually increased the speed of the centrifuges causing them to destroy themselves. In the event, the Iranian scientists realised something was wrong and the damage was limited to about 1,000 centrifuges, but even this number seriously set back the Iranian uranium enrichment capacity. Unfortunately, during the attack the virus escaped onto the internet and infected millions of computers worldwide, but only caused damage if they carried the particular Siemens software targeted in the Natanz plant.

The ability to launch cyber attacks are in their infancy. For instance, they often rely on a human vulnerability to introduce a piece of software to a system, as above in the Natanz example. This somewhat limits your ability to launch attacks—reportedly the North Korean nuclear plants were subject to a Stuxnet operation but attempts had to be abandoned because it proved impossible to introduce the virus into the target computer systems. Already under development are a class of artificial intelligence-enabled cyber attacks, which automate the ability to penetrate and attack target computer systems—effectively making them more effective, but much more unpredictable.

Tactical cyber operations

Tactical scale cyber operations are not simply smaller versions of strategic scale cyber operations. To begin with, at the tactical level, military computer systems are more likely to be the target, and these are generally hardened and encrypted making them much more difficult to attack (ensure your military systems are hardened and encrypted otherwise they will be shut down by a moderately capable enemy).

Therefore, at a tactical level, offensive cyber operations will look for the weakest link. In the twenty-first century that is most likely to be the mobile phone that every soldier carries.

Gaining access to them will allow you to listen to everything they say, track where they are, and photograph anything in their camera's lens. You can also gain access to all their contacts, allowing you to play mind games with their relatives, degrading their morale (note: take mobile phones away from your soldiers, and punish them if they are found with one). In Afghanistan and Iraq, all British soldiers had their phones taken from them and returned at the end of their tours.

There will be other weaknesses that you can exploit and/or guard against. Many militaries use small commercial drones for simple reconnaissance tasks like observing enemy troops. Although cheap and effective, they are easy to hack into and reprogramme. This idea extends to almost anything commercial and unencrypted that your soldiers may have with them on the battlefield: sports watches, exercise trackers, tablets, etc. All of them need to be taken from your soldiers if you want to ensure full cyber security.

As with at strategic level, tactical-level cyber attacks are likely to be effective for only a short duration, until your enemy realises what is happening and rectifies the problem. This means that if you can find a means of penetrating key enemy systems—like air defence, command and control, or battlefield data-sharing networks that help guide weapons to their targets—then you should only use them at the point of an important assault or manoeuvre. If your enemy remains unaware of the software vulnerability that you plan to target, then you can compound their confusion and perhaps divert their attention, while you conduct other movements or attacks on the battlefield.

Looking to your own cyber defence, you should always assume that your critical systems may have vulnerabilities and could be shut down, much like you should always assume that your communications may be being intercepted. For this reason you might consider training your troops on how to fight successfully if key electronic systems stop working (when it 'goes dark'). Do they know how to navigate with paper map and compass? Can your ships and aircraft navigate without computer systems? How will your troops communicate if your radios stop working? Can your supply systems still deliver fuel and ammunition if they only have paper and pencils to organise it with? Most professional militaries are already reverting to including these methods in their training, and you should too.

Tactical information operations

Tactical information operations—that is, delivery of information to enemy troops, or to civilian populations in your operational areas—is hard to do, and generally not very effective. Traditionally, militaries have relied upon leaflets, posters and radio broadcasts to get their message across in so-called psychological operations. The process of generating these messages relies on developing an understanding of what enemy troops or the civilian population might be thinking or feeling during that phase of the war. This is extremely hard to do well, and so messages tend to be blunt instruments ('surrender now and you will be treated well'; 'we are here in peace to help develop your country').

The simple reality is that enemy troops, even if they hate their chain of command, are more likely to believe them than they are enemy propaganda directed at them. And force or the threat of force is overwhelmingly more persuasive than leaflets. Enemy civilians are usually in a similar position, and most civilians are simply trying to survive the conflict and are unlikely to be swayed in their thinking by a leaflet.

There are some more subtle types of tactical information operation that target individuals, and these are generally kept very secret by those militaries that have the capability to carry them out. But, if you are very skilful, and have an excellent knowledge of the enemy command structure, you can 'spoof' (fake) conversations between enemy commanders on, for example, their mobile phones where each commander thinks that he is communicating with the other, but in fact they are both communicating with you.

This allows you to start cultivating distrust or sow operational confusion. A simpler version involves starting a private conversation with a key enemy commander to convince him that if he surrenders, for example, he and his family will be allowed safe passage to a new life somewhere idyllic. These types of operation are very hard to do well—as they rely on intimate knowledge of the enemy's key personalities—but they can be among the most successful type of tactical information operation.

However absent this highly intimate knowledge, tactical information operations are a blunt instrument. If possible, you should maintain the ability to conduct them (radio transmitters, printing presses for leaflets) but not pin too many hopes on their effectiveness.

8.

NUCLEAR, CHEMICAL AND BIOLOGICAL WEAPONS

Often lumped together, nuclear, chemical and biological weapons constitute quite different things. This chapter will set out how you should think about them and how their existence should change your behaviour.

Nuclear weapons, with the ability to wipe out whole cities in one strike, and the probabilities of mutually assured destruction, change the fundamental calculus of war. Almost everyone agrees that this level of civilian destruction is unacceptable, and therefore there are strong international norms in place against their use. Nuclear weapons serve as big red lines—geographical, or about intervention or escalation in a particular war—which your enemies ought not cross.

Chemical weapons, on the other hand, have significant tactical uses—for example, flushing defenders out of underground defensive positions—but they are indiscriminate and also kill civilians. Unfortunately, norms against their use have never been robust, and they have seen recent use in the Syrian civil war. This, coupled with their great effectiveness as a weapon in urban combat, probably mean that you will see increasing use of chemical weapons in the coming decades.

Biological weapons are different still. Most experts agree that biological weapons are simply too difficult to manufacture, store and deploy on a large scale—and using them can harm your own troops and civilians as much as the enemy's. In short, they are generally overrated as a large-scale military threat, although recent advances in biomedical science may mean that we are entering a new era of their use.

Nuclear weapons

Nuclear weapons can generate a scale of destruction that is unmatched in human history. By harnessing a nuclear reaction, they generate huge amounts of explosive power from relatively small amounts of matter. And although tested extensively (usually in deserts, remote atolls or under the ocean), nuclear weapons have only been used twice.

At the end of the Second World War the United States dropped two bombs on the Japanese cities of Hiroshima and Nagasaki, thus forcing Japan to surrender. The explosions were each equivalent to between 15,000 and 22,000 tonnes (15–22KT) of TNT (normal explosive), despite the devices themselves only weighing 4.5 tonnes each. At least 200,000 Japanese civilians and military personnel died, many of them slowly and painfully as a result of radiation sickness, and large areas of the cities were flattened. Not surprisingly, their use was, and remains, extremely controversial.

As terrifying as the destructive power of those bombs was, modern nuclear weapons can be several orders of magnitude more powerful. The most powerful nuclear weapon ever was tested by the Soviet Union in 1961. Nicknamed the 'Tsar Bomba', the weapon delivered an explosion that was equivalent to between 50–58 million tonnes (50–58MT) of TNT—approximately 2–3,000 times the explosive power of the bombs dropped on Japan. The bomb was so powerful—it created a 5-mile-wide fireball, and shattered windows 560 miles away—that it had to be deployed with a parachute to slow its descent so that the deploying aircraft had enough time to escape.

The awesome power of nuclear weapons creates stability in world affairs—at least between the states that have them. The idea is that no nuclear power will attack another with nuclear weapons because the retaliation would result in mutually assured destruction, known as MAD. Extending this idea, it means that when tensions do arise between nuclear powers, there is always a keen understanding that escalation must stop at some point, and it pushes these countries to set boundaries to their disputes and conflicts.

In other words, nuclear weapons create a psychological structure to the international security environment. And to operate at this level, almost all nuclear powers seek to maintain a nuclear weapons capability known as a minimum credible deterrent—that is, enough nuclear weapons (and delivery systems) to respond to a nuclear attack against you with enough force to, say, destroy your attacker's capital city.

Coupled with minimum credible deterrence, most nuclear weapons states maintain ambiguity about their exact nuclear posture—that is, exactly when and how they would use nuclear weapons—so that their adversaries remain in the dark about how they would respond to an attack. This enhances the stability of the system because a nuclear aggressor cannot be sure that their adversary won't attack them back, thus enhancing the concept of mutually assured destruction.

Nowadays nuclear weapons tend to come within roughly three size brackets. The largest are between 400–800KT, and the most common are those of about 100KT. These weapons are still extraordinarily destructive: the blast radius of a 100KT nuclear weapon is approximately 2km, within which buildings will be flattened and anything flammable incinerated.This is then followed by the irradiation of the landscape through the falling of radioactive dust ('nuclear fallout') which continues to kill people in the weeks, months and years after the weapon's use.

You will probably hear much about so-called 'tactical' or 'battlefield' nuclear weapons, with a low yield in the 1–10KT range (for comparison the bombs dropped on Japan were 15KT and 22KT). These are designed to be used on critical enemy infrastructure like bridges, command posts, or logistics dumps. Because of their lower yield, it is suggested in some quarters that their use is less consequential, and so they are more likely to be used. This is not the case. All nuclear weapons are nuclear weapons, and because of the ambiguity around the posture of nuclear states, it is not at all clear that the use of a 'tactical' nuclear weapon would not result in a retaliatory nuclear attack of a much larger scale. With nuclear weapons, all routes lead back to mutually assured destruction.

In 2023, there are nine known countries with nuclear weapons. Five of them are the permanent members of the United Nations Security Council: China, Russia, France, the United Kingdom and the United States. In addition, Pakistan, India, Israel, and North Korea have all tested nuclear weapons, and keep them for use. Russia and the United States have by far the most—a relic of the Cold War that neither power has managed to shed—with around 5-6,000 each (and a mixture of different warhead sizes). China has approximately 350 (sizes unknown), and the UK and France have 2–300 (with most of their warheads in the 100KT range).

There are several weapons delivery systems in use beyond the original method of simply dropping nuclear weapons from a plane. Many countries have land-based intercontinental ballistic missiles (ICBMs) either stored in missile silos or on portable launchers that are constantly on the move. Many countries carry their nuclear weapons on submarines from which ICBMs can be fired. Exactly which size of warhead and which method of delivery a country adopts depends on what it perceives to be its greatest threat and is based on planning to avoid all its nuclear weapons being destroyed in a surprise attack (known as a 'guaranteed second strike').

Russia and the United States have developed and maintain for use all three types of delivery system: land, air and sea. To a large extent this is a relic of the Cold War, as the large numbers of nuclear weapons and delivery systems available to them are enough to level every square inch of both countries to the ground and destroy the global ecosystem most likely causing genocide of the human species. The UK and France use nuclear missile submarines (and aircraft in the case of France) to maintain what they argue offer them 'minimum credible deterrence'. That is, both countries expect to survive a nuclear strike against them, and to be able to respond: they have guaranteed second strike.

Traditionally, China has kept a small number of land-based missiles which enabled it easily to strike India, or Russia. However, due to the 11,000km from China to the continental United States, and the ability of the latter to destroy some ICBMs in flight, it is less clear that China would could carry out a guaranteed second strike against the US. Therefore, and because of the increasing levels of strategic competition between China and the US during the 2010s and 2020s, China has been modernising and expanding its nuclear forces to give it the same nuclear triad that the US and Russia have: land-, air- and submarine-launched nuclear weapons, thus guaranteeing a second strike for China against any potential US attack.

The picture is complicated by nuclear proliferation to other countries. That is, if many countries end up with nuclear weapons, or arms races occur in particular regions between competitors, then the possibility of their use, or an accident occurring, increases. This was probably put best by former US Secretary of State George Shultz who spoke of the sense of dread around nuclear weapons receding as more countries possessed them.

And the picture is yet further complicated by anti-ballistic missile defence: if a country were able to guarantee that it could shoot down every incoming missile (or plane), then it could destroy the enemy's systems without them being able to respond.

As yet, no country has a complete missile defence. Russia, for instance, has a shield around Moscow. And the US has a more extensive system across the continental US, and in some US possessions in the Western Pacific like Guam, but the system is only capable of defending against a missile attack from a less technologically advanced country like North Korea, rather than a full-blown attack from Russia. Similarly, China has some sort of limited missile defence, but little is known about its extent or capabilities. Certainly at the moment, the big nuclear powers are all vulnerable to each other, thus maintaining mutually assured destruction.

This stable, yet fragile, nuclear balance has so far held. Over the decades it has been reinforced by arms control treaties—mostly between Russia and the US. These have limited the numbers of nuclear weapons held, limited short-range missiles (which increase instability because there is no warning of their launch), and limited anti-ballistic missile defence (because successful missile defence would reduce the possibility of mutually assured restriction).

In the 2020s, we are on the cusp of a more unstable world. The least of this instability is being caused by China upgrading its nuclear deterrence forces. More important is the expiration of or withdrawal from various arms control treaties by the United States under the Trump administration, thus encouraging Russia to develop new types of hypersonic missiles to deliver their nuclear weapons (which have a Moscow-Washington flight time of fifteen minutes, and so reduce the warning time). And most important of all for this increasing instability is the potential for many lesser powers—like Iran and Saudi Arabia—to consider acquiring nuclear weapons.

If you lead a country that possesses nuclear weapons, keep hold of them. They are your ultimate insurance policy, and they enable you to lay down certain red lines in defence of your interests that you would not otherwise be able to do. In so far as you are able, you should seek to work with other nuclear-weapon-possessing states to mutually reduce your stockpiles, as this reduces the chances of nuclear accidents. If you are a very large nuclear power (United States, Russia, and soon-to-be China) then you should work towards arms control agreements as these increase stability in the international security system.

But there is something that you must accept about nuclear weapons: because they are so destructive, the psychology that surrounds their use is completely different from the use of normal weapons.

For a start, in a conventional war, you use your weapons to inflict maximum violence on your enemy. Conversely, the aim with nuclear weapons is not to use them. Therefore, how you communicate with your enemy is completely different. It is vital to be understood clearly, and while there might be deliberate ambiguity under which precise circumstance you would use nuclear weapons, your enemies must be in no doubt that you would do so if you feel the circumstances justify them.

Normally, there will also be complete clarity about certain aspects of your policy: for example, that you will respond to a nuclear weapons attack on your territory with a nuclear attack on their territory. The concept of deceiving your enemy—which is ubiquitous and important in conventional warfare—is not relevant in nuclear deterrence. This is because deception about intentions—or perceptions of deception—reduces nuclear stability. You or your enemy do not want to feel that the ability to launch a guaranteed second strike is being eroded.

If you are not a nuclear state, you must think very carefully about whether you want to acquire nuclear weapons. You might decide, for instance, that you are under enduring threat of invasion, and that nuclear weapons are the only way of dissuading your would-be attacker from invading you (Israel and North Korea would argue that this is their position).

You may fear that your regional competitors—with whom you may have had several wars—have nuclear weapons, and that this would give them the edge in any future conflict (India and Pakistan, and several competing powers in the Middle East find themselves trapped in this logic). Or you may decide that you are a great power, and great powers need nuclear weapons to maintain their position (this was the position of France, the USSR, Britain and China when they acquired their nuclear weapons in the decades after America used them at the end of the Second World War).

The technological hurdles to producing a nuclear weapon are not that significant. And unless you plan to deliver your weapon via a truck into your enemy's capital city, the delivery system—usually a long-range missile—will offer a similar level of technical challenge to your scientists. The problems in acquiring nuclear weapons are almost entirely political ones: other countries will not want you to get hold of them, and in all likelihood you will have signed away your right to develop nuclear weapons in the Non Proliferation Treaty.

The NPT—which came into force in 1970—is the most widely signed arms control agreement in history with 191 signatory countries. At its heart is a bargain: the nuclear weapons states (defined as China, USSR/Russia, the US, the UK and France) agree not to share nuclear weapons technology with other countries, but to share the peaceful benefits of nuclear power. They also agree to work towards eventual nuclear disarmament. In return, the non-nuclear weapons states agreed not to pursue nuclear weapons. This bargain held with only three non-signatories at the time (India, Pakistan and Israel) and one withdrawal (North Korea after their nuclear weapons test).

The reality painted in the Non Proliferation Treaty is that the major nuclear weapons powers do not want other countries to acquire nuclear weapons because it will destabilise the existing delicate nuclear balance. And they will initiate major foreign policy actions, including extensive economic sanctions and even the threat of war, to stop other states acquiring nuclear weapons.

If you feel that you must acquire nuclear weapons for one of the reasons outlined above, then you must pursue the objective as secretly as possible. Even then, it will be hard to keep it hidden from one of the major nuclear weapons states who all have effective intelligence operations. Therefore, it would be prudent to assume that you will be found out, and that pressure—most likely extensive economic sanctions, although as we saw in the case of Iran that pressure may include cyber attacks—will be applied to your country. There is also another subtlety in that pursuing nuclear weapons may enable you to gain concessions out of great powers if you are to verifiably denounce your nuclear weapons programme (for instance security guarantees or preferential trade deals). That said, however, if you manage to reach the stage of possessing a functional and credible nuclear weapons deterrent, you have largely guaranteed that your country will not be invaded by another power.

You must weigh up whether the pursuit of nuclear weapons will benefit you strategically. If you manage to acquire them, you will possess the ultimate insurance policy; but if you fail, you will suffer strategic losses that stem from exclusion from the international system. You may ultimately conclude—and it is your judgement alone as leader of your country—that the pursuit of nuclear weapons may not be in your interests as much as you originally thought.

Chemical weapons

Nuclear and chemical weapons are often grouped together under the rubric of 'weapons of mass destruction', but their similarities end at the fact that they both cause excessive suffering to civilians who will inevitably be seriously injured in attacks on military targets.

The first large-scale battlefield use of chemical weapons was during the First World War when all combatants used them—unsuccessfully—in order to try to break the deadlock caused by the opposing trench systems. It has been estimated that there were 1.3 million causalities of the gases used which mostly caused disfigurement and injury through blindness, choking, and blistering of the skin (although not necessarily death). Included in this figure are 260,000 civilian casualties.

During the Second World War, chemical weapons were used extensively by the Japanese, especially against Chinese troops. And although not a battlefield use, the Nazi regime used different gases in the Holocaust's gas chambers that killed over 3 million people, most of whom were Jewish. This is the greatest death toll from the use of chemical weapons. Since the Second World War, such a large scale of chemical weapons usage has ceased, although 100,000 Iranian troops were victims of Iraqi chemical weapons during the Iran-Iraq war of the 1980s.

Chemical weapons come in four varieties: blister agents, nerve agents, blood agents and choking agents. Blister agents—like mustard gas—cause large fluid-filled blisters on the skin, eyes and respiratory tracts. Nerve agents—like Sarin—affect the nervous system and cause convulsions, paralysis and death (if they affect the heart and/or respiratory system). Blood agents—like cyanide—work by preventing the uptake of the oxygen carried in the blood. And choking agents—like chlorine gas—work by causing a build up of fluid in the lungs which inhibits breathing.

Nerve, blood and choking agents can kill extremely quickly in the right concentrations, although at lower levels will 'merely' injure horrifically. Blister agents take longer to kill (typically up to a day) and are more likely to result in injury rather than death. All four varieties of chemical weapons tend to be either liquids or gels at ambient temperature and the most common delivery method is artillery shell.

The most complicated aspect of their delivery is that once you have deployed them you cannot move into the territory that you have attacked until the chemical has dissipated or unless you are willing to send your troops in with protective equipment. There is also the problem—prevalent since the First World War, where it happened quite often—of the wind direction changing, thus creating the danger that you gas your own troops.

The use, development, production, stockpiling and transfer of chemical weapons is banned by 193 countries under the 1993 Chemical Weapons Convention. At the time of the signing of the Convention, several countries including Russia and the United States had stockpiles of chemical weapons, which have almost all been safely disposed of at the time of writing in 2022 (the US, with one of the largest stockpiles, will have completed the task by the end of 2023). In 2017, Russia declared that it had disposed of its chemical weapons, although the use of the Novichok nerve agent in several assassinations (or attempted assassinations) brings into question whether Russia has completely destroyed its previously held weapons.

There is one exception to the Chemical Weapons Convention that many argue should be brought under its purview: the use of white phosphorous. White phosphorous is used to create very thick smoke very quickly, and is delivered by hand grenade, vehicle smoke tube, mortar or artillery shell. It is by far the most effective screening agent—it can even confuse or confound thermal imaging systems—enabling you to move troops and vehicles unseen. It is particularly used when troops find themselves ambushed and need to withdraw quickly without their exact movements being tracked by the enemy. But white phosphorous can cause burn injuries to the skin, and also to the respiratory tract if inhaled.

The distinction that is drawn under the Convention is how white phosphorous is used. If used for screening purposes (or sometimes for marking things on a battlefield to, for example, direct aircraft to a target) then white phosphorous is considered legal. However, if it is used directly onto enemy positions to cause harm to an enemy it is considered to be a use of a chemical weapon under the Convention. As you can see the distinction is a subtle one that is very hard to police in battlefield conditions, so be quite clear with your military about how white phosphorous is to be used.

It is a reasonable assumption that although there are almost certainly small quantities of chemical weapons held by various states around the world, large-scale stockpiles have been destroyed. And these small quantities are still sometimes used on the battlefield against either enemy soldiers or civilians—witness the recent use of (mostly) chlorine gas by the Syrian armed forces against anti-government rebels and, inevitably, civilians.

Unfortunately, despite being barbaric, chemical weapons do have military uses particularly in urban warfare: for example, many of the gasses are heavier than air and are therefore perfect for killing defenders that are dug in to the ground, or who are using cellars, metro systems, or other underground excavations for cover. They also—understandably—create complete panic when they are deployed causing defensive lines to collapse. Finally, they do have uses in denying whole areas of the battlefield to an enemy (like a river valley), which forces them to take the routes that you want them to take—like through a minefield.

This military utility, particularly in urban areas, coupled with other factors like their ease of manufacture and delivery, and their sporadic use throughout the years resulting in a lack of substantial norms against their use, leads to the conclusion that we are likely to see the continued resort to chemical weapons—albeit at a small scale.

Therefore, it is prudent that you prepare your troops to survive a small-scale chemical weapons attack. For relatively little expense you can issue every soldier with a plastic or rubber chemical suit and a respirator. It is also not that difficult—or expensive—to train them how to decontaminate themselves and their equipment after a chemical attack. With these simple steps, you will ensure that most of your troops survive a chemical attack. You may wish to also invest in a small number of chemical weapon specialists who can analyse the weapons used against you, to help your defence in the future. All in all, chemical defence is not a very expensive or difficult investment.

Should you develop or use chemical weapons? Ultimately, their use must be balanced up against the opprobrium you will receive from other countries, including the so-called 'great powers' (Syrian government facilities were attacked by the US, UK and France after the use of chemical weapons in the Syrian civil war). So if you in any way need the consent of other nations, then the small tactical advantage that you will gain from the use of chemical weapons is likely to be hugely outweighed by the strategic disutility accrued from loss of consent. It is instructive to note that the Great Powers have not used chemical weapons for many decades, largely because at some level they all seek the consent of other nations, which the military use of chemical weapons would erode.

Biological weapons

The first recorded use of biological weapons was in approximately 1500–1200 BCE when victims of tularaemia (a bacterium) were driven into enemy lands causing an epidemic. The Mongols catapulted bubonic plague victims into besieged cities. And in the 1700s, British soldiers deliberately introduced smallpox into populations of native American Indians and aboriginal Australians who were attacking them. Biological weapons have been around a long time, and are considered both tactical weapons (that you would use to target enemy troop formations) as well as strategic weapons (that you would use on, for example, an enemy capital city).

In modern times, the only systematic, large-scale use of biological weapons was by the Japanese on Chinese military and civilian targets during the Second World War. In one example, ceramic 'bombs' containing fleas infected with the plague were dropped on a city. In another, large quantities of paratyphoid and anthrax were spread amongst food and water supplies as the Japanese withdrew from an area. In this instance, around 10,000 Japanese soldiers infected themselves and around 1,700 died, thus illustrating the key difficulty of harnessing biological organisms for military use—they are very hard to control once you have released them.

Biological weapons are usually bacteria—small single-celled organisms—and occasionally viruses, or biological toxins (poisons). Over the twentieth century several militaries, including those of the United States, the United Kingdom and the Soviet Union developed extensive biological weapons programmes where multiple organisms like the bubonic plague, anthrax and tularaemia were weaponised (mostly by being sprayed from aircraft in droplets of water). They all have incubation periods of around three to five days and untreated death rates of between

60–80 per cent. Most militaries view anthrax as the most serious biological weapon, because although vaccines are available, they are not habitually given to troops and a course of vaccines over eighteen months is required to develop immunity in 90 per cent of vaccinated people.

The ideal biological weapon is one that can be delivered either by dropping the organism, or its spores, on the target area. Again ideally you want a bacteria that causes the disease directly, and does not create secondary person-to-person infection, otherwise you risk setting off an epidemic that you cannot control, and which you are just as likely to be a victim of. You would also want to have the ability to vaccinate your own troops against your own biological agents, thus minimising the risk of blowback.

This leads to a key problem in the use of biological weapons: if you use a bacterium or virus that is highly lethal, it is likely to do as much damage to you as to the enemy; yet if you choose an organism that you can vaccinate against or otherwise control the spread of, your enemy is likely to have the same capability.

It is this problem, and the reason that biological weapons usually take several days to make their targets ill (the organism's incubation time), that has meant that militaries have concluded that they are too difficult to use on a widespread scale. They are more trouble than they are worth.

Similar to the other 'weapons of mass destruction' discussed in this chapter, the use of biological weapons—particularly against civilian populations with all the attendant risks of unintended spread—would be considered unacceptable by most people and nation states. It is highly likely that any country using biological weapons would immediately become an international pariah and forfeit any consent for their war aims. (This conclusion does not stand for terrorist groups for whom the incubation period offers them an advantage in escape time, and who care less about collateral damage, blowback, or consent.)

The impediments facing military use of biological weapons meant that 184 states signed the 1975 Biological Weapons Convention. Similar to the later-signed Chemical Weapons Convention, the Biological Weapons Convention bans the use, development, production, stockpiling and transfer of biological weapons. It seems to have established and reinforced strong international norms against their use. (Again, this Convention clearly doesn't apply to non-state actors or terrorist groups who are the most likely actors to use biological weapons.)

There is an open question, however, as to whether we will again see the use of biological weapons. Even setting aside terrorist groups or death cults, which at the moment is probably the greatest risk of their use, there are several developing technologies which enhance the utility of bioweapons to nation states.

First and foremost among these is the development of cheap and fast gene editing. In theory, and as far as I am aware, no-one has done this, yet gene editing of a virus or bacteria could enhance the properties (or mitigate the weaknesses) of an organism, thus making it a more effective weapon. For example, gene editing could enable a nation state to reduce the incubation period of an organism, while increasing its lethality.

The current pace of technological change in biomedical science is so rapid that it will change medicine completely over the next thirty years. In addition to the fairly mundane changes outlined above, these developments will open up new and terrifying possibilities in bioweapons. It is not beyond the realms of possibility that we will one day have the capability to 'design' an organism that targets a particular ethnic group, or people who eat a particular food. Or perhaps the organism—anthrax, for example—could be edited so that it can evade current vaccines, while at the same time a working vaccine could be developed for your own troops.

Alternatively, an organism with a long incubation period, and a relatively low death rate—similar to Covid-19—could be designed to shut a society down over an extended period of time, enabling an adversary to attack. The possibilities are only bounded by our ingenuity.

* * *

Nuclear, Chemical and Biological weapons represent a unique category of threat that you should treat differently to the 'conventional' methods of war-making outlined elsewhere in this book. Nuclear weapons, if you possess them, require a very careful consideration of your strategic signalling to adversaries. The most important thing is for the nuclear weapons that you possess to contribute to overall international security stability by using them to draw red lines above which conflicts do not escalate. Because of the seventy-five years of thinking that have gone into nuclear stability and the widespread terror of mutually assured destruction—barring accidents or proliferation—this 'nuclear peace' should hold.

This is not the case for chemical or biological weapons, sadly. Chemical weapons have real military uses and can take immediate effect over a relatively small area. For these reasons, and because full humanity-wide norms against chemical weapons have never held for more than a few years (that is, they haven't held), it is likely that chemical weapons will continue to be used, albeit on a small scale.

Biological weapons present yet another case where the norms of non-use established under less developed levels of technology may come under increasing strain as nation states seek to gain competitive advantage against other nation states. Even so, there are huge problems with the use of even 'engineered' bioweapons, and so many nation states will be extremely cautious about using them.

These simple truths about biological and chemical weapons explain why you must train and equip your troops to survive small-scale attacks—protective suits and respirators at a minimum, and the ability to decontaminate themselves and their equipment so that they can continue fighting. Your troops should be vaccinated against the obvious biological agents like anthrax. But you should not use them, for the reasons outlined earlier in this chapter.

Part 3

HOW TO FIGHT A WAR

9.

THE ART OF USING LETHAL VIOLENCE

In this chapter I seek to answer the question discussed throughout this book: how do you fight a war on land? Here I will describe generalship and the art of conducting the orchestra of war to demonstrate how to use the building blocks set out in the preceding chapters: both the intangible foundations you have built up—strategy, logistics, morale, and training—and the tangible capabilities that you will have developed—in the land, sea, air and space and cyber and information domains—to defeat an enemy army in the field. This chapter will describe when you should use certain capabilities, and why (with the exception of nuclear, chemical and biological warfare).

Strategy, narrative and morale

The first things to consider are the strategic goals you have set yourself, and the wider narrative that you seek to reinforce by using lethal violence. Are you seeking to punish an enemy government for bad behaviour towards you and hoping to convince the wider world of the just nature of your cause? Perhaps you seek to liberate a people from the tyranny? Are you defending your own home territory or overseas possessions? Are you part of an international coalition seeking to enforce a United Nations mandate to protect civilians from genocide? Are you annexing a neighbouring country because you believe that it should not be an independent state?

Although you will have decided your strategy and accompanying narrative well before you launch any military action, the precise type of war you are fighting—liberation, annexation, enforcement, defence, peacekeeping, etc.—will to some degree set boundaries on how ruthless you are in the application of lethal violence.

These boundaries will not only be formally set out in the orders that your generals follow, but are important in setting the narrative for your own soldiers as they go into battle—they must have the clearest idea of what you are seeking to achieve, their role in it, and how they must act in their pursuit of victory.

Imagine, say, that you have a pocket of tens of thousands of enemy soldiers trapped. Whether you seek to kill every last one of them, or force them to surrender, or create circumstances that encourage them to run away will depend very much how you wish your use of violence to be interpreted.

The wider world will most likely forgive you killing tens of thousands of trapped enemy soldiers if you are defending your home territory; this is much less likely if you are on a UN peace enforcement action, where it would probably be preferable to force their surrender.

A key example of this occurred during the First Gulf War. In 1991, having forced Iraqi troops from Kuwait City, the US-led coalition had the option of striking thousands of Iraqi soldiers as they retreated north up Highway 80 towards southern Iraq. An attack was ordered which destroyed approximately 2,000 Iraqi military vehicles and killed around 1,000 Iraqi troops (although no-one knows exactly how many Iraqis died). The event is still controversial and many people in the Middle East contend that the soldiers were retreating and should not have been attacked, although the Americans maintain—rightly under the laws of war—that they were engaging legitimate military targets who had invaded and remained in another country.

There is a final reason why your information operations narrative—for that is what this is—must be thought out at this time, namely its impact on your own troops. You should carefully consider why you are asking your soldiers to risk their lives, and those of their friends, and as far as possible this has to match what they experience on the ground, especially when they start taking casualties.

If they have been told that they are to liberate a country from an autocratic and evil regime, but when entering its towns and villages ordinary people resist and attack them, you will begin to experience flagging morale and a reduction in their will to fight.

You are also probably storing up long-term mental health problems for your troops because of the dissonance between what they are told and what they experience, and because that dissonance is experienced within the highly stressful environment of combat where lives are lost, and people are horrifically injured. I know people who suffer acute mental health problems because of this dissonance of narrative that they experienced while serving with British forces in Afghanistan.

These ideas also translate into how you conduct your battle in order to affect the psychology of your enemy—and particularly your opposite number, the enemy commander. This directing of the battle—known as operational art—is strongly based on a shrewd understanding of human nature. It is very much an 'art', and your judgement as commander is what makes the difference between success and failure.

As a commander, always root your leadership in psychology, and base your actions on what will have the greatest dislocating effect on your opposite enemy commander. The options available to you as a ground commander are only limited by your imagination.

Exactly what this looks like is in your judgement and experience: sometimes destroying an enemy elite unit will affect the morale of the whole army; sometimes bypassing combat units and striking at logistics will cause the enemy army to collapse; sometimes fixing enemy units with artillery while you seek to destroy their command and control will overload their decision-making. But you must always think about your end goal: the psychological collapse of your enemy, and specifically the enemy commander.

Everything you do on the battlefield is part of a narrative story that is partly told through your physical 'military' actions and partly through your information operations. Humans understand events—and particularly wars which are complex—through stories. It may sound extraordinary and macabre that you will risk your soldiers' lives, and seek to kill countless enemy soldiers, simply to tell a story, but this is the difference between commanders that win wars, and commanders that lose wars.

If you, as the overall commander of your forces, can tell a story that your troops, the enemy troops and commander, and the wider world population can understand, then you will be a successful military leader.

Let us say that you wish to tell a story that your army is all-powerful and will keep moving forward no matter what, and the enemy will be crushed irrespective of what they do, and that therefore resistance is futile and they might as well give up. You might call this the 'Mongolian Hordes' narrative.

One way of telling this story is to conduct big, very powerful attacks, with lots of artillery and aviation, and massed troops supported by tanks (hopefully) overrunning the enemy positions. Your military actions will give you the backdrop to tell this story of overwhelming strength through the media (think of images of massed rockets firing into the air) and in your meetings with other world leaders.

A good demonstration of this occurred when British forces intervened in Sierra Leone in 2000 in support of the government against a rebel group, the RUF. At all times, the British sought to send the message that they were an overwhelming military power using helicopters, fast jets, and highly-trained paratroopers in order to convey the stark notion that there was no point in fighting them. They were successful in defeating the rebel group and laying the foundations for a peace that, at time of writing over twenty years later, still endures.

Alternatively, you may wish to tell a story that you are nimble and clever, and your opponent is a lumbering behemoth that cannot respond quickly enough—the 'David and Goliath' narrative. One way of telling this would be to try to break through your enemy's lines at one point and get behind them, causing havoc. Alternatively, you could hit your enemy hard in one area, and then switch to another axis and attack them there creating the impression that it is you dictating the pace of the battle and your slower, lumbering enemy is having to respond.

Hezbollah did this very well in their conflict with Israel in 2006 where they continued firing rockets at targets in Israel, despite massive Israeli military operations to deter them from doing so. Eventually the UN brokered a ceasefire, but many outside observers consider the war an 'information victory' for Hezbollah (and it led to a wide-ranging internal review by the Israeli armed forces to attempt to learn the lessons of the conflict).

There are of course other stories you can tell, but you should always try wherever possible to tell one that everyone can understand. Doing so gives your own troops a narrative to be part of, and if successful you impose a story on the enemy's troops and commanders that they are part of without wanting to be.

More broadly, the wider world audience will understand your battle, and your victory in terms of your narrative. And if you can use your battlefield activities to reinforce the overall narratives with which you are trying to paint the conflict—your information operations as described in Chapter 7—then you will win the war.

How big should your operational force be?

Once you have set a strategy and narrative that the wider world, and your own population and military, can work with and believe in, you must consider some of the simple mathematics of land warfare. These numbers are immovable constraints to fighting on land—if you don't have enough troops you will fail, for instance—and they are often the facts 'wished away' by overconfident political leaders. You should not make these mistakes.

At the very highest level, you should look at force ratios—the number of your troops compared to the number of your adversary's troops. This stems from the simple premise that it is much easier to defend than to attack. The general rule of thumb, which assumes that levels of technology are comparable, is that you will need three times as many attackers as you have defenders opposing you (3:1). This ratio can raise to 5:1, or even 10:1, in complex urban environments. Of course, the ratios will be different if one side has artillery or air supremacy and the other does not.

As you advance, you will need ever increasing numbers of troops, because you must garrison the enemy settlements and lines of communication (roads, rail, bridges, etc.) in your rear areas. This is extremely manpower intensive. Occupying an unfriendly town could require a tenth of the population in troops. Maintaining your force ratio as you advance becomes exceedingly difficult as you will be constantly balancing between protecting your supply lines (which will stop your operations if they are severed) and maintaining enough troops at the front lines with the enemy.

You can see from these rough calculations that one very quickly gets to an army size in the hundreds of thousands. This is before you take into account your wounded, killed, or simply exhausted soldiers—all of whom will need to be replaced. In addition, you should consider maintaining a strategic reserve—up to a third of your force—to deal with the inevitable unforeseen eventualities that arise in war.

During the Cold War, both NATO and the Warsaw Pact (the Soviet Union's alliance system) had about 3 million troops each in Europe. Thus neither side had anything like the force ratios required to overwhelm the other in conventional warfare. This is one of the reasons that nuclear weapons took on such prominence during the Cold War.

Not having enough troops will force you to make sub-optimal choices on the battlefield—leaving your supplies unguarded, leaving a flank thinly protected, or relying on artillery to make gains when you should be using tanks and infantry—which will ultimately expose you to the enemy more and lead to you taking more casualties. This problem repeatedly hampered Russia during its 2022 invasion of Ukraine. Attacking with only 150,000–200,000 troops on multiple fronts, a small fraction of what was required, the Russians failed to generate high enough force ratios to overwhelm the Ukrainian defenders.

You may be able to mitigate lack of troops in limited areas and temporarily by concentrating your forces in another location to achieve 'overmatch', and then moving your forces elsewhere to repeat the trick. Your enemies will also be trying to do this to you, and so clever use of sequencing and deception is paramount. Overall, though, not having enough troops can end up as a downwards spiral as you will be forced to take risks with and exhaust the ones you have.

These calculations offer a simple way to see whether your force can achieve its aims. They cannot be wished away, and if you don't have enough troops, you should reconsider the scope of your operations, otherwise there is a strong possibility you will fail. And once you have settled on the size of your army, your industrial base and trading system must produce and deliver enough fuel, armaments, and equipment to your troops fighting in the field.

What capabilities do you need?

Once you have looked at the size of your army, the next step is to ensure that you have the right capabilities and trained troops to defeat your enemy.

First, look to the terrain that you must cross: if there are multiple rivers for instance, then you will need lots of bridging equipment. Can your army conduct mountain warfare? How do you intend to cross the soft bog that surrounds the enemy's capital city with your armoured vehicles? You must think about these issues before you advance across the start line, always considering: 'Is there an element of the terrain that will stop my operations, and if so, how do I mitigate against that factor?'

Second, you must consider your enemy's capabilities. First and foremost, do they have air power? If you are attacking them, if at all possible you must first gain air supremacy—this means not only destroying their aircraft, but also any air defence assets that they may have, and degrading their airfields so that they cannot be used (you may not seek to destroy them permanently, so you can use them later—one option is to use cluster munitions to crater the runways and taxiways, which is difficult but not impossible to repair).

Studying your enemy's air capabilities can become incredibly technical. For example, Russian air-to-ground missiles are designed to target the specific bandwidths used by NATO radars. This became a huge problem when invading Ukraine because the Ukrainians originally relied on Russian-designed air defence systems, later supplemented by Western ones. You really must think very carefully about who you will be fighting with, and often many years in advance.

If you cannot destroy your enemy's air capabilities then you must ensure that you can contest them in the air with your own assets or (more likely) by equipping your troops with abundant air defence missiles (man-portable air defence missiles are quite cheap and are easy to distribute among your troops, although you will need larger, vehicle-based systems for defending against missile attack). If they have copious cheap and small drones, can your anti-air systems target them? If not, you need that capability before you arrive in the field.

Next, you must look carefully at the capabilities and balance of their ground forces, namely between their infantry, artillery and tanks—the three types of forces that form the nucleus of any land army. In essence, this stage of the analysis is about understanding what the enemy's army can do and ensuring that you can counter it.

First, artillery: What are the maximum ranges of their artillery versus yours? If there is a great mismatch—that is, their maximum range is 100km and yours is 30km—then you must either get longer-range artillery, or some method of neutralising their artillery like, say, armed drones or helicopters. Otherwise, the enemy will be able to hold your infantry off at a distance and, more importantly, will be able to hit your logistics without you being able to do anything about it. This single factor could easily cause your operations to grind to a halt.

Second, balance: Is your enemy force unbalanced? Are there too many tanks compared to infantry? Confirm you have enough anti-tank systems dispersed among your troops. Too much infantry without tank and artillery protection? Check you have enough artillery with the right types of ammunition to deal with massed infantry. Too much artillery compared to tanks and infantry? Consider carefully what capabilities you will need to restrict the ammunition supply to their artillery.

Third, you must consider any specialist capabilities that your enemy might have and ensure that you have a counter. Perhaps they can lay very effective minefields, or you may be concerned that they have chemical weapons. Each element of the enemy fighting force must be analysed, and your force shaped accordingly.

Mission planning and deception

By now, you have the right size of army for the enemy you wish to defeat. It is comprised of the right capabilities in the correct proportions to defeat them. Now is the time to plan and issue orders to your troops.

The first thing to focus on is the psychological effect that you wish to have on your enemy, and particularly on your enemy commander. Partly, this will depend on the type of army that you have—if you can only field a poorly-trained conscript army then your only option may be to fight a war of attrition where you force your enemy to retreat or die (attrition is a notoriously slow form of warfare that only works if you have overwhelmingly more troops than your enemy, and is usually only practised by poorly trained conscript armies because they are unable to do more complex operations). North Korea, with its million-plus army, is prepared to fight attritional warfare.

A more highly trained force, with modern equipment, particularly aircraft or drones, has more options and flexibility in conducting variations on manoeuvre warfare where it seeks to strike the enemy's critical functions (usually their command posts, communications, and logistics) to destroy their ability to fight. This then forces the enemy commanders or government to withdraw or surrender and enables you to win without killing every enemy soldier.

Once you have decided upon this, you will need to prioritise which of your enemy formations you wish to target. Let us say that your overall goal is to occupy your enemy's capital city and to install your own administration. Standing in your way are twenty-five enemy divisions (some quarter of a million troops). Which you attack first illustrates a key concept in military planning: sequencing versus concurrency. Namely, do you attack them all at the same time, or do you attack them in sequence.

If your logistics allow, attacking a number of enemy formations simultaneously brings with it several advantages. Most importantly, it is likely to overload their logistics and ability to deal with casualties. It will also mean that rarer assets—such as rocket artillery, helicopters, or reconnaissance and surveillance troops—can only be used by some formations and not others. But most importantly, attacking concurrently will overload your enemy's ability to make decisions quickly, particularly if they rely upon a hierarchical style of command, where all major decisions must be made at the top of the command structure.

However, it is much more likely that you will be operating under the same restraints constraints as your enemy and must consider sequencing your attacks. Even though you are being forced to do so you should still always try to overload your enemy's decision-making ability, hampering their ability to defend themselves. The classic way of doing this, used by humans since fighting in clans of hunter-gatherers, is to surprise your enemy or deceive them as to your intentions. There are as many ways to do this as there are to launch an attack, but broadly deception breaks down into two categories: ambiguity increasing deception, and ambiguity decreasing deception (see Fig. 7).

Both categories rely on a central fact about war: that it is complex and ambiguous (the so-called 'fog of war'), and you have no option but to make decisions based on insufficient information. If you wish to increase the enemy's feelings of ambiguity, then you must present multiple options to them of what you might do. Because they don't know what your actual plans are, their decision-making will be paralysed. If you wish to decrease your enemy's feelings of ambiguity, then you must convince them that you intend one course of action, when in reality you have a completely different plan in mind. All deception in war is an artful blend of these two concepts.

Both ambiguity increasing and decreasing deception rely on the same repertoire of tactical options (of which there are an almost infinite set of variations). Well-known methods include the feint, where an attack is carried out in one location, only for the main attack to fall in another area; a demonstration, where troops are massed in one area, only for the main attack to fall somewhere else; and a ruse, where false information is fed to the enemy, such that they expect a development that is not your plan.

At the tactical level, this deception might involve manufacturing wooden 'units' or fake radio nets. At its most simple, it could involve building up forces in one place, only to attack in another, or staging a 'retreat', only to counterattack hard. These particular tricks have been used for as long as humans have been fighting: as every British schoolchild knows, Norman troops at the Battle of Hastings in 1066 repeatedly feigned retreats to draw the English down from their higher defensive positions, whence they were killed. The Normans won, and the British Isles has never looked the same since.

Because of the internet and particularly social media, we now have very effective ways of disseminating information—part of your battlefield shaping could be demonstrating your military activity in one area while imposing a communications blackout on another, thus drawing your enemy's attention to the former area.

You can also use social media to spread fear and panic among enemy troops if you can generate enough images of their destroyed equipment and dead soldiers. In 2022, Ukrainian forces generated and dispersed many videos of destroyed Russian vehicles during a breakthrough of Russian lines in the Kharkiv Region in the country's north-east. This turned a breakthrough into a rout as Russian morale collapsed and many of their soldiers abandoned their positions and equipment.

As at the strategic level, surprising your enemy, and deceiving them at the tactical level, can be much quicker routes to victory than trying to smash your way through in an attritional style of warfare.

However, surprise on the battlefield is easier said than done. When you do attack your main target—known as your main effort—you must use overwhelming force to achieve victory. In other words, you must 'Clout, not Dribble' (in the words of the British Royal Armoured Corps). But even then, armoured units—tanks, armoured infantry, and artillery—can only advance around 30km per day once you take into account logistics and maintenance (and that's without delays predicated on heavy fighting).

Simple maths will tell you how long it will take to get to your objective, and this time lag, coupled with ubiquitous modern sensors like drones and satellites, mean that it is very hard to gather enough force together in one location—and use it—without being discovered (and frustrated in your plans).

Logistics, again, battlespace management, and orders

It is at the mission planning stage that you must—once again—consider logistics. You might have the most daring, bold plan to manoeuvre around your enemy's formations and strike at their rear. But if you are unable to see how to supply your troops once they are at your enemy's rear then you must think of another plan.

The hard reality of logistical planning—and the enormous volume of stuff that you must transport—means that the logistical requirements of your army will most probably dictate your tactical plan. For instance, some of your initial objectives probably will not be enemy troops but bridges, roads and rail yards. Your tanks and infantry might have off-road fighting capability, but the remainder of your army will be constrained to the road network—and so most of the fighting will occur close to main arteries.

Once you have a plan that is logistically supportable, divide up the area where you will conduct your operations—called the 'battlespace' in military parlance—and assign it to your formations and units. Different units (and their commanders) will have boundaries between them over which they cannot cross, or fire weapons, without permission from the unit that 'owns' that battlespace. It makes sense also to identify your enemy's boundaries between their units, and attack across the boundaries or down the seams—this makes it harder to coordinate a response.

This battlespace management is primarily important in stopping you mistakenly attacking your own units, but also to ensure that you do not frustrate your neighbouring commander's plans (maybe they were waiting till the enemy concentrated in one location so that they could destroy them all at the same time) (see Fig. 8).

You should also remember that battlespace management is three-dimensional. The airspace above your ground troops will also be very contested and not only by aircraft, helicopters and drones—every time your ground troops fire mortars, artillery or missiles through the airspace it must be de-conflicted so that you don't shoot down one of your own planes by mistake.

Once you have the foregoing questions answered, it is time to issue your orders. If you are an army that is more hierarchical, then your senior commanders will have planned in detail how the smallest groupings of your soldiers will act. The best advice suggests that you adopt a 'mission command' style of leadership, where each formation informs its subordinate formations of the overall plan, tasks them with their objectives, but does not go into too much detail about how these are to be achieved. Your orders should then cascade from the top of your army down to the bottom—from army, to corps, to division, brigade, and battalion, all the way down to company, platoon and section—with each level thinking about their own part in the plan, and how they can best achieve it.

Control of the air

You will find it extremely hard to degrade your enemy's leadership, communications, logistics, infrastructure and key assets, as well as channelling them effectively, unless you have some degree of air supremacy. Complete air supremacy means that you can fly your air assets anywhere you like, including over territory that enemy ground forces control. When the French were deepening their involvement in Côte d'Ivoire in 2004, the first thing they did was destroy the country's tiny air force, before moving in 300 ground troops.

To achieve control of the air, you need to do several things. First, you must target your enemy's air defence systems, preferably by missile strike. But you should also degrade their radar systems, and other elements of their military (and civilian, so they cannot repurpose it) air traffic control systems. There will also likely be plenty of man-portable, shoulder-launched air defence systems (known as MANPADS). MANPADS have a maximum altitude range of approximately 20,000ft, so once you have destroyed most of their air defences, your planes will be safe if they fly high enough.

Next, you should target your enemy's air assets like planes, helicopters and drones which can attack your troops or logistics. This usually occurs at the same time as you degrade or destroy their air infrastructure—airfields, runways, taxiways, aircraft hangers, and aviation logistics like fuel storage tanks (storage tanks are usually underground and so you will need bunker-busting munitions).

Here you have a decision to make: do you want completely to destroy the enemy's aviation infrastructure so that it can never be used again, or do you intend to take control of their airfields and other facilities in the hope of adding them to your assets later on in the campaign? If the latter, then you will still want to destroy their air defence systems, but only crater their runways with cluster munitions, it being a relatively simple task to bring them back into service at a later date.

Shaping the battlefield

Now you can start to deploy lethal violence. Initially, this should be done in what are called shaping operations. This is a phase of the battle, operation or war where you restrict your enemy's future choices. The classic way of doing this is to target and destroy your enemy's leadership elements, communication networks, logistics, and key infrastructure. Attacking leadership and destroying headquarters begins to degrade the enemy's ability to make decisions. Attacking their means of communications impedes their understanding of what is going on and scope to direct the battle (see Fig. 9).

At this stage you are likely to be using air power especially including missiles, or special forces raids (you are only ever going to use special forces for very high-value, symbolic or strategic-level enemy targets), and offensive cyber operations. These 'strike' operations should be combined with information operations to misdirect your enemy as to what you are doing.

If possible, you should not only focus on battlefield communications, but attempt to destroy, jam or misdirect their strategic communications—if your enemy is a powerful country this will inevitably mean attacking their satellite constellations and undersea fibre optic cables. As you take out higher-end capabilities, the enemy should be forced onto other means of communications that are hopefully easier to intercept, and less secure. Ideally, you want to destroy enough communications architecture that the enemy is forced to resort to mobile telephones or unencrypted radios to keep in touch. In Afghanistan, NATO forces sometimes blocked the mobile phone networks to force Taliban fighters onto simple radios and satellite phones, both of which were easier to track.

Taken together, attacking command, control and communications (known as C3) atomises the enemy force, breaking down a corps into divisions, and divisions into brigades, and so on. It also begins to erode the morale of the enemy force as their understanding of what is happening becomes more and more ambiguous, and their ability to respond becomes increasingly tenuous (also known as getting inside their OODA loop [see Fig. 10]). Humans—and particularly soldiers—like certainty, and if you take that certainty away from them, life becomes a lot more stressful, and they become much less effective.

At the same time as disrupting and destroying your enemy's C3, you must pay attention to their logistics. As will be clear to you by now, an enemy force without fuel or ammunition is simply an array of expensive targets for you to deal with at your leisure. Fuel particularly will have an obvious distribution network starting at the refineries. If these are within the battlespace, and you are content that they will significantly degrade your enemy's ability to fight (rather than just shutting off the fuel supply to their civilians), you should destroy their refineries.

If the fuel is refined elsewhere and transported in, then target and destroy their fuel distribution network, starting at the back—the oil-receiving terminals at the docks—and moving forwards towards the front line. At this stage, a few well-placed cruise missile strikes, or special forces attacks, on the enemy's main oil-importing route could literally bring their army to a standstill.

Ammunition supply may be harder to target as it is likely to be dispersed and hidden. If you can find it, hit it, but otherwise you can destroy the enemy's factories, so that when their current stocks run out they will have no reserves to replace it. At this stage you may decide that it is easier to target key pieces of infrastructure which will stop the ammunition and other supplies getting to where they are needed.

Bridges across wide rivers are the most obvious target. They are pinpoint targets whose location is well known (as opposed to docks or rail yards which can be dispersed over a wide area). They are also easy to damage with precision-guided munitions so that they cannot be used. If there are no bridges, then you should look at railway yards, major road junctions, military depots, truck yards, and so on. In short, target any infrastructure that stops your enemy getting what they need—but above all ammunition and fuel—to their fighting troops.

What you are trying to do in this first shaping phase of the battle is to degrade some of the Intangible Fundamentals upon which your enemy's combat power rests. Attacking leadership and communications affects their intelligence understanding and ability to coordinate their strategy or plans (Chapter 1) as well as their morale (Chapter 3). Attacking supplies and infrastructure affects their logistics (Chapter 2).

Thus the shaping of the battlefield is a psychological shaping, to put your enemy on the back foot, to force them to deal with the problems you have created for them, and limiting their ability to attack you.

Shaping too is about generating momentum—probably the most important psychological phenomenon that you want on your side—and making your enemy respond to you, rather than the other way around. This feeling of momentum that you have, and the dis-momentum felt by your enemy, will be reinforced as your enemy's ability to make decisions becomes degraded or overloaded, hopefully causing a downward spiral and collapse in their will to fight.

The aim of any battle is to convince your enemy to stop fighting and accept defeat (this is because it is much, much easier than killing them all). The reality is that troops that have no leaders, cannot speak to each other, are running out of food, and ammunition will either become ineffective, a much easier target for you to attack, or will abandon the fight because of poor morale. The Israelis successfully accomplished this in the Six Day War in 1967 against several adversaries including Egypt, Jordan, Syria and other Arab states. The initial attack destroyed most of the Egyptian Air Force, and targeted communications systems. This was coupled with a ground assault against Egyptian positions in the Gaza Strip and the Sinai Peninsula. So dislocating was the speed of the assault that Jordan and Syria barely joined the war, and all sides sued for peace within a week, leaving Israel with a significant victory.

Beyond destroying the psychological fundamentals of your enemy's combat power, there are two other things that you might wish to consider targeting during the shaping phase of your operation. The first is called channelling: restricting your enemy's ability to use certain parts of the battlefield, thus forcing them to use other routes (other 'channels'). This can be obtained by forcing them to use certain river crossings, supply routes, sea lanes (by dropping parachute mines in the sea lanes you want to block them from, for example), and airfields.

Second, your enemy will likely have key capabilities or specialist equipment or troops that enable them to conduct particular military manoeuvres. For example, they might have specialist radar that enables them to trace your artillery shells while in flight giving away the positions of your guns. Other examples include mobile bridging equipment that enables them to cross rivers, minefield clearance vehicles, or satellite constellations.

What you are looking for in this phase are any assets that your enemy has that are (a) rare, (b) hard to replace because they are, for example, very expensive, or (c) are crewed by personnel that take a very long time to train (in which case the personnel are the target). Removing or disabling enough of these assets from the battlefield will generate a permanent loss of capability for your enemy because they will be unable to regenerate them before the war ends, or without significant cost.

All of the tasks of battlefield shaping—destroying or disrupting the enemy leadership, communications, logistics, and key infrastructure and assets—and indeed the tasks of fighting a modern battle require a systemic approach to understanding the enemy through intelligence, and then prioritising the targets, and attacking them. Militaries call this targeting and those militaries that have an efficient targeting process, or 'kill chain', tend to win battles.

The targeting process and intelligence fusion

The act of prioritising which parts of the enemy military machine to attack with your own scarce military resources is ancient. But an efficient, data-driven, continuous process that reacts to intelligence in almost real time has only been possible with the advances in communications, sensors and information technology of the last few decades. Among Western militaries, this process was refined and developed particularly by special forces operators during the first two decades of the twenty-first century, and then adopted by the wider 'conventional' armies.

The targeting process has a wide applicability not just in the shaping of the battlefield but also as the actual battle begins (known as the decisive phase). Ultimately the problem of too many enemy targets and not enough of your own resources to attack them all almost always exist. The targeting process solves this problem by helping you understand which targets are more important to the enemy (and hence to you), and then directing appropriate military assets onto them.

The more efficiently you can conduct this process, the quicker you are able to destroy enemy targets. This—particularly if you are targeting leadership and communications—enables you to psychologically disorient your enemy, as you start to make decisions faster than they do. This is known as getting inside their decision-making loop and herein lies the way to collapse their cohesion and unravel their military power.

A targeting process consists of four stages (see Fig. 11). First, there is a process of 'finding' (understanding) the enemy targets. This is done through a process of intelligence 'fusion', described below. This will ultimately give you a list of targets and the priority with which you should attack them. You may also decide that some targets you wish to destroy, some you wish to disable, some you wish to capture, and so on.

Intelligence fusion is about blending different sources of intelligence. As you will recall from Chapter 1, there are several methods of gaining intelligence: from human sources, communications or signals intelligence, electronic signatures, surveillance images, observations of your own troops, and other highly technical and specific sensors like infrared and air turbulence. Fusion is the process of bringing all these different sources together to illustrate what you know about a particular object on the battlefield.

In addition to more traditional types of intelligence, most military intelligence organisations will have a team that searches and analyses information from social media—particularly Telegram channels (a messaging app that allows large groups to share information) and Twitter (where eyewitness information is blended by information collators and analysts).

In recent wars, where in some armies soldiers carry their own mobile phones, record videos of their experiences, and post extensively on social media, finding targets can be much easier than in the past. In some cases, such is the wealth of social media data produced, it is possible, through geolocated videos and posts for example, to track the movement of a military formation across the battlespace.

It should not be underestimated how important this is for understanding the battlespace, and how detrimental it can be to armies that have poor communications security. The Syrian civil war—ongoing from 2011—is the conflict where this came of age, with all sides using social media to post, tweet, and broadcast, often inadvertently giving away important targeting information in the process.

This type of battlefield intelligence gathering is conducted at a much faster pace than the strategic intelligence gathering that was described in Chapter 1. In an ideal scenario, you will understand and prioritise your targets in a matter of minutes, or at the very least hours. Strategic intelligence gathering—which is often not linked to direct actions but conducted for background understanding—works on a scale of weeks and months.

Your enemy will know that you are attempting to target their leadership, communications and logistics and will have hidden or camouflaged their headquarters, protected their communications networks, or kept them moving so that they are more difficult to target. They will have also tried to protect their logistics (or not left them static for long) and key pieces of terrain or infrastructure like major bridges (they will not want to be channelled down certain routes, instead seeking to retain their freedom of manoeuvre across the battlespace). To mitigate against these defences, you must 'Fix' the enemy targets in place—this means keeping them in the same place, or at least under surveillance so that you know of any changes to their location or defensive posture (see Fig. 12).

Third, you 'Finish' the target which is military parlance for destroying it (or capturing it, etc.). To do this, your targeting team will have to 'bid' up the chain of command for a strike asset making clear the priority of the target that you wish to destroy. Because there are always more targets than there are assets to destroy them, and there may be other people across your forces who need that fast jet bombing run, you must have a central mechanism for allocating strike assets to different targets.

Finally, you must conduct a further piece of intelligence work to understand whether the enemy or battlefield has changed now that you have destroyed that target. Among the military this is known as Exploiting and Analysing. The whole process is known as Find, Fix, Finish, Exploit, Analyse, or to the military, who love acronyms, F3EA. And like so much of what makes militaries effective, it is a rigorous process, efficiently executed.

What does targeting look like in real life?

Let's say, for instance, that you are interested in targeting what looks like an enemy headquarters building.

The headquarters was originally spotted by signals intelligence that identified a particular location as a hub for radio signals spreading over a wide area. The density of signals leads you to believe that it is a divisional headquarters, but you are not sure. From a human source, you learn that a General John Smith has been seen in that area in the last week—was he based there, or merely paying a visit? Your open-source team report that photos of Col Dave Jones leading a patrol in the nearby town have surfaced on Twitter—you know from other intelligence that General John and Col Dave have a mentor-mentee relationship. Full generals don't normally lead divisions though, so what is going on?

You task satellites to take some images, and the layout of the tents and vehicles looks too big for a divisional headquarters, but it nevertheless seems to be an important location of some sort. Finally, you order an analysis in the electromagnetic spectrum, looking at heat, among other emissions, which tells you the amount of electric power being produced and consumed at the location. This confirms that it is a headquarters, but is it an enemy corps headquarters—a target of the highest priority that you want to destroy? It is only through fusing all these different types of intelligence together that you have been able to understand in such detail what the target is, and hence where it sits on your priority list.

Next you must keep your target under surveillance. For a smaller target, you might loiter a drone over it, but one of this importance is likely to have air defence assets arranged around it. You decide to task a reconnaissance satellite to keep it under observation, and to task your signals intelligence capability to continue monitoring their transmissions volume—any change in these may indicate that the headquarters is moving, or that something else is about to occur (you are unable to understand the contents of the transmission due to encryption, but the volume of transmissions and the directions from which the replies come is still very useful information).

Next you must cue up the assets to destroy the headquarters. You may have several options available: bombing from the air, a special forces raid, or a missile strike from one of your submarines off the coast. You decide that the risk of losing a plane to the enemy's air defence is too great, and a special forces raid too risky, leaving you with a multiple cruise missile strike from your submarine. For maximum effect, you decide to time this strike to be five minutes before you launch an armoured thrust through the enemy front lines in the sector controlled by the corps headquarters that is your target. For good measure, you launch a cyber attack on their radar systems at the same time.

Once the strike has gone in, and your attack is underway, you must continue to monitor the target in the Exploit and Analyse phases. Let's assume that you have managed to destroy the enemy headquarters. As a priority you should listen across the entire enemy front for radio signals—you have just taken out a major node in the enemy's communications network.

You will want to look for several things. Where does a node reform in the network to provide command and control to the enemy forces? Does one of their divisional headquarters step up and become the corps headquarters? Is there a backup corps headquarters? Does the enemy lose the ability to function at that particular headquarters' level, and begin to prosecute the battle in this sector as a collection of separate smaller formations? All these observations will help you line up the next set of targets in your targeting process.

Although a simple example, you can see how this process allows a military to find and understand potential targets across a battlespace. Inevitably there will be more targets than you can strike, and this process allows you to prioritise them, and decide which of your capabilities you are going to use for which target. As the battle unfolds, you can coordinate attacking high-value targets like headquarters and communications elements. This system, or one like it, can be used to assign a target number (and priority) to every enemy target—every tank and sniper—and to prioritise your military capabilities onto them.

Your enemy—if they are professional—will be targeting you as well. So you must think hard about how to hide or protect these critical elements from enemy attack, or deceive your adversary as to which are important. Additionally, each of you will be seeking to attack each other's targeting processes and kill chains. This duelling of the two kill chains, both against battlefield targets, and against each other, is known as hiding versus finding, and the military that finds the quickest, and hides the longest, will win the battle.

Increasingly, this process is becoming more and more automated, in order to speed it up. Each stage benefits from automation and algorithms. The vast array of intelligence data can be very effectively analysed allowing patterns to be found by artificial intelligence (a large part of intelligence analysis is pattern recognition). Targets can be prioritised and assigned much quicker by computer than by human. And—not yet, but potentially in the future—the decision to Finish targets could be generated by an artificial intelligence system. We will look at the future of automation and artificial intelligence in the Epilogue.

Operational art in the advance

Once you have shaped the battlefield, decide what you are going to do about each particular grouping of enemy forces. Some you may choose to destroy; others you may wish to engage to keep them occupied and unable to disengage (to 'fix' them in place). Yet others you may choose to bypass, either to surround them, or to get behind them and destroy their logistics, thus rendering them ineffective.

As you conduct this battle you must keep your enemy guessing as to your priorities, and where destruction will be wrought on their forces. Your preparations—above all your logistics—must be as far as possible hidden from view. Your troop movements—particularly the reinforcement of certain positions to enable an advance—must also be kept from the enemy (or what look like similar reinforcements elsewhere made to keep them off-balance).

Then come the operational questions. Where are you going to destroy? Where are you going to fix their forces? And where are you going to bypass? Do you destroy the major bridge at the enemy's rear so that they are bottled up, or do you deliberately leave them an escape route? Perhaps you could damage it so that heavy equipment and supplies cannot cross? At this stage your air force and your navy are likely to be operating in support of your ground priorities and attacking targets that will help your ground manoeuvre. If you have any offensive cyber capabilities now may be the time to deploy them—are there any critical enemy systems or networks that you can incapacitate even if only for a short time?

Ultimately, one of the aims of this manoeuvring is to create local 3:1 force ratios that enable you to attack successfully. Let us say, for instance, that you and the enemy have forces that are completely matched in size. Both sides will be trying to create situations where they have a 3:1 ratio in a local area, so that they can prosecute a successful attack, before regrouping their forces to push forward an attack in another area.

One of the most critical decisions that you have to make as leader is when to commit your forces to the decisive battle—when do your tanks, infantry and artillery move en masse and seek battle with the enemy? In warfare, timing is everything. The judgement that you have to make is whether your shaping operations have sufficiently limited your enemy's ability to respond, or at least their range of responses, to your attack.

Will they, for instance, be trapped in a certain piece of terrain because you have destroyed the bridges? Will they be unable to coordinate a response because you have destroyed their headquarters network? Are they unable to bring forward enough ammunition and fuel to defend or counterattack? Or have you convinced them through deception that your attack will fall in one place, only for it to fall in another?

China is clearly practising shaping when it intermittently blockades Taiwan. In the event of a conflict, a successful blockade of the island would significantly limit the options that the Americans (and others) have in supporting Taiwan. More broadly, China is also shaping the entire South China Sea region by building new islands, reefs and atolls to position its forces deep into maritime space previously thought to be international or belonging to other countries like the Philippines.

Once you have decided to launch a decisive offensive your forces are committed. This is it. There are no half-measures. You must inflict maximum violence where you have decided to inflict it. You also must—as with the rest of the recommendations in this book—understand and remember that the violence you are inflicting is designed to have a psychological effect. As in the shaping phases, you are still likely to want to prioritise leadership, communications, logistics, and rare enemy capabilities, although you may choose also to strike enemy tanks and infantry in order to grind down their combat power.

Destroying enemy combat troops reduces their ability to respond to you but also makes them feel weak and unprotected (and hence more likely to run away). However, your aims are not necessarily to kill every enemy soldier, but to win the battle in the minds of the adversary. If that involves bypassing large numbers of enemy soldiers on your way to strike important logistics targets in their rear, then so much the better. If this bold move pays off, you are also probably likely to lose less of your own soldiers this way.

As well as inflicting maximum violence, you must seek to maintain momentum. Momentum is not the same as going as fast as you can—your tanks and armoured or mechanised will outrun your logistics tail almost immediately if you do. Momentum in warfare means making decisions faster than your enemy—if you can continue to do this then you will be constantly making them respond to your actions, rather than you responding to theirs. This way you will win your battles and, quite possibly, the war.

There are as many ways to maintain momentum as there are types of rifle. The most obvious is to advance rapidly through enemy-held territory or, if you have enough troops, to advance on multiple axes. The latter manoeuvre allows you to choose where and when to strike, giving you space and time in which to make decisions that your enemy must respond to. In turn it means the enemy must decide on which axis it wants to use its rare assets—like air defence.

Another option is to launch an attack on one axis to get your enemy to commit some of their forces to defence, thus allowing you an option to strike in another area. You can destroy leadership or jam communications elements to slow down your enemy decision-making—remember, to maintain momentum, the only goal is to make decisions faster than them.

Once you have worked out how to maintain momentum, you must think about keeping your troops fresh. The 24/7 intensity of modern warfare is such that humans cannot maintain peak effectiveness for very long. You must build in rest periods for your troops, otherwise your army will rapidly fall apart. This is best done in sequence, where once they have conducted an assault, or fought a battle in the defensive, you are able to take them out of the line and replace them with fresh troops.

And so, in the advance, it is best to conduct a 'leapfrogging' type of movement where one unit conducts an attack and sits in defence on the enemy objective once taken. A second unit then passes through their position (known as a 'passage of lines') and continues the assault. Ideally, you would have three such units in rotation with one in the assault, one acting to suppress the enemy (to keep them busy while the assault troops manoeuvre into position), and one resting or in reserve. (You will notice that this one-thirds rule keeps cropping up in combat: you need three times as many troops to attack as defend; there are three platoons in a company, three companies in a battalion etc.; assault-suppress-reserve is in thirds; etc.)

In addition to this, keeping some of your force back as an operational or strategic reserve is essential. A reserve allows you to respond to any unforeseen enemy activities—the enemy has a vote in your success or otherwise. Even more importantly, an operational or strategic reserve enables you to exploit any successes that you may have. Once again, timing is everything.

Once you have decided when to commit your troops to the decisive phase, the next most important—and most difficult—judgement that you must make is when to commit your reserve. The Japanese did this well in the Arakan Campaign in Burma in 1942. They waited until the British offensive had run out of steam, and then deployed their reserve, which was fresh, to push the British back.

It is likely that you will make this decision without the luxury of full factual awareness of the situation confronting you. By now, the 'fog of war' will have fully descended. Confusion about where the enemy is and their strengths and intentions will abound. It is likely that at any one time you will not know the full status of your own units. In assessing whether to deploy your reserve you must assess whether the enemy assault that is about to break through your lines is the real thing or merely a deception.

In respect of your own attacks and assaults, if you collapse your enemy's front lines, and get your mechanised or armoured columns behind them, will deploying your reserve at this critical moment enable you to break the enemy's defence across the whole sector of the front? Amid this ambiguity you must also decide on something that is very difficult to reverse: once committed, it is hard to disengage and regenerate a military reserve.

Manoeuvring around the battlefield

Both before, and once you have engaged with and assaulted the enemy, you will still need to manoeuvre effectively across the battlefield. In some cases this is fairly simple: advancing across open countryside entails adopting different formations depending upon the exact topography, type of the terrain, and what you understand of the enemy threat. For instance, when moving across open plains you might decide to adopt an arrowhead formation with your main body either directly behind the arrowhead or arranged to your flanks to enable you to deploy your force quickly to the enemy's flank.

If you are forced to move along a road—perhaps you need to secure it as a logistics route, or it is the only route through a forested area—then inevitably you will need to adopt an extended column. Clearly, in this formation you are at risk of being ambushed by the enemy, and so you must think about protecting your flanks either by having your vehicle turrets facing sideways or by deploying small packets of soldiers onto your flanks, or well in advance of your main body, to find the enemy before they can ambush you.

No matter the type of terrain or enemy threat there is always an advantage to occupying high ground. High ground allows you to look down on everything, making it much easier to see the enemy and to direct your fire onto them. Conversely, if you occupy a hilltop then it will be much harder for the enemy to direct fire on your troops and assault your positions. This simple military fact has been true since our primate ancestors fought for prime position on the best fruit tree in the forest.

Beyond general manoeuvring, which must be practised and carefully choreographed, there are certain 'set-piece' movements which are exceptionally difficult. The toughest of all to execute are river crossings and retreating.

In a river crossing—particularly a contested one, or one where there is a high chance of the enemy interrupting your crossing—you will be vulnerable to ambush or attack and find it difficult to respond. Consider what must be done to transport an armoured infantry brigade (of 3,000 personnel and 300 tanks and supply vehicles) across a river (see Fig. 13). The selection of your crossing point is vital—ideally not too wide, nor too fast (river bends are probably out), and if you are lucky a place to cross where you have high ground on the near bank so your forces can be over-watched.

First, you will need to reconnoitre the near bank—preferably well before with drones and confirming the picture you get with infantry on foot. You will also want to get the best possible understanding of any potential enemy dispositions on the far bank. If you have the reconnaissance assets, you will need to look well beyond the far bank into likely avenues of approach in order to detect any enemy troops moving towards your crossing point. Next, you should establish heavy weapons—your tanks, mortars, artillery—in a position on the near bank, or on high ground if there is any, where they can observe and direct fire onto the far bank.

Once you are confident that you are in a defensible position, you must put a vanguard over the river. If you are sure there are no enemy around, then you may choose to skip this step and go straight to deploying your bridging equipment over the river. But if you decide that you do need to establish yourself on the far bank, you should only involve a small number of troops—perhaps 50 or 100—swimming across the river with their equipment. This is because if it goes wrong and your troops get ambushed, only a company's worth of troops will be in the enemy's kill zone. Once on the far side, these troops need to conduct further reconnaissance of the far bank, check for obstacles or traps, and then establish a defensive position forward of the crossing site.

You are now ready to bring forward your bridging equipment—perhaps an automatic extendable type on the back of a single lorry for a small crossing, or a constructible pontoon type for a larger crossing.

As your engineers start to construct the bridge, you should be ready for enemy artillery or air attack, and have ready your own artillery to fire back at the enemy artillery batteries (I will describe the exact method below). Ideally, if you have the resources available, you would have your own air cover up—either helicopter or aircraft—in order to respond to any enemy moves at this moment of your maximum vulnerability.

Assuming you get the bridge in place without being disrupted by the enemy, you should immediately send over a mixture of tanks and armoured infantry. Once they are on the far bank, they should push outwards from the crossing point to establish proper defensive positions, relieving your earlier vanguard of dismounted infantry. You can now use the crossing to get large numbers of troops across and build up your forces on the far bank.

As you can see, manoeuvring troops safely when confronted by an enemy threat is complex, and involves as much skill as fighting. Yet a river crossing is not the most complex military operation you will encounter. Retreating your troops in good order without your forces collapsing is far more difficult. Probably the most skilfully executed one of all time is the 1,000-mile retreat of British forces in the face of a Japanese offensive in Burma during the first half of 1942. The British, although losing 10,000 men, managed to keep the lion's share of their army (comprising mostly Indian troops) alive, and delayed the Japanese advance sufficiently so that defences could be prepared in India. These lines held, and enabled a British counterattack, which eventually retook Burma.

Retreats do not necessarily mean that you are losing the war—you may decide to retreat to deceive your enemy, or to reinforce another axis of advance. But the fact that retreats are usually conducted to stave off defeat and the annihilation of your forces make them fiendishly tough to accomplish.

In a retreat, you face several problems. First and foremost, there is the psychological element. Whether your troops are retreating to avoid being encircled and killed, or whether they are executing a pre-planned battlefield movement, soldiers like to go forwards, and not backwards, so you must as a priority maintain morale and offensive spirit.

The second problem is logistical. You must ensure your front-line forces have enough supplies to conduct the fighting withdrawal, yet you do not want too much fuel, ammunition and food forward because you are about to de-occupy the territory and hand it to the enemy (you can destroy your supplies rather than backload them, but that is hardly ideal). You will therefore need to pay great attention to logistical management as your troops retreat.

The third problem is choreographic. Probably the easiest way to conduct a retreat is what is called a rearwards passage of lines (see Fig. 14). This is where you have set up parallel lines of your troops, and the one nearest the enemy disengages with the adversary and passes back through the second closest line of your troops facing the enemy, thus allowing them to take on the fight. This is difficult to do without shooting your own soldiers as they pass through your own lines.

These three problems interlock and make it very difficult to retreat without it turning into a rout, where your troops run away and are slaughtered. This is even more the case if you are retreating under duress, where your soldiery is likely to be exhausted, injured and psychologically worn down. Then the temptation to down arms and run away is strong, and only the most highly trained troops with the best morale can execute this move well under fire without disintegrating.

Defending positions

Now we come to defending positions. Defence is a vital element in your toolkit as a commander. It enables you to tie up large numbers of enemy troops and conduct vital preparations behind your defensive lines. Defence is the mechanism through which you stop the enemy forces and fix them, in preparation for a thrust from another direction. It enables you to fight attritional defences where you give ground and positions very, very slowly while inflicting maximum damage on the advancing enemy. The Ukrainians have used this tactic during the 2022 Russian invasion of their country to grind down and inflict heavy damage on the enemy, in return for small gains in territory. But, most obviously, if you are the party being attacked, it enables you to stop losing territory, infrastructure and population to the enemy.

The selection of defensive positions is paramount, and in many ways has not changed since pre-historic times. Elevated positions like hilltops, plateaus, and ridges are favoured. Alternatively, you can position yourself so that you have water features—rivers, streams, and marshland—in between you and your opponent. A combination of the two is best.

If your defensive terrain is short of topographical or hydrological features, then you may wish to consider using urban areas for defence. Not only do they have symbolic value as centres of population and administration, they are usually road and rail communications hubs, which you can use for your logistics, denying them to the enemy. But most important of all—they give a huge advantage to the defender because they create a three-dimensional landscape of buildings and underground areas.

If your enemy wishes to assault a city that you hold, they will need to consider gathering forces of up to ten times the number of defenders that you have in place. Urban defence is a great way of tying down large numbers of enemy soldiers. Of course, you must weigh up this stark military advantage with the fact that the city will likely be full of non-combatants, and civilian casualties are notoriously high in urban warfare. Probably the greatest urban battle of all time was fought in Stalingrad during the Second World War. The Soviets managed to halt the German offensive in the city, whilst preparing another formation that swept through the German lines outside of the urban area, thus trapping significant numbers of German troops in the city. It was a major turning point of the Second World War.

Once you have selected your locations for defence, your engineers should build you defensive works—tank ditches, bunds (raised earth barriers), and trench systems—either to enhance areas that are promising for defence, or to create them from scratch. Failing that, your soldiers should dig into the ground with their own entrenching tools, which every soldier carries. All obstacles that you create or take advantage of—from rivers to dug ditches—must be covered by your own troops or artillery, otherwise the enemy can cross them at their leisure. Obstacles not covered by fire are not obstacles!

Establishing defence positions and lines is a key activity in the advance and the retreat. It enables you to rest your soldiers, reorganise them, repair equipment, for your logistics elements to catch up with the fight—all known under the rubric of an 'operational pause'. Indeed, you may wish to pause and defend specifically to set up a logistics or repair area behind your lines, thereby preparing your force for future operations. Alternatively, you may wish to hold terrain to switch your efforts to another axis of advance.

Finally, in a fluid situation, where your troops have just made a large advance through the enemy lines, there will come a point where you will want to defend again for the above reasons. Often when this happens both sides will skirmish and jockey to establish the best positions for what they probably hope is a temporary defensive position.

To conduct a successful defence, you should base your plans on six principles (see Fig. 15). First, all your defences should be layered—what is known as 'in depth'. This means that as well as a front line of positions facing the enemy, you have two, three or more further lines of defence behind them. So if you lose one of your front-line positions, the enemy cannot break through your lines and start destroying your logistics or command and control elements.

Similarly, you want to know that you can defend in every direction, so that if the enemy decides to flank you, all of your troops are not facing in the wrong direction—this is known as 'all-round defence'. To this principle, I would add extensive surveillance and reconnaissance capabilities (if you have them) thereby enabling you to assess from which direction the enemy is likely to come.

Third, each of your defensive positions should be in a position of mutual support that allows them to give covering fire to at least one other defensive position (and preferably many more than one). In this way all your positions support each other, or in the event they are overrun, attack each other!

Fourth, you must keep a reserve of around a quarter of your troops. The enemy will have successes which you must react to, and they may leave themselves vulnerable in the assault, which opens up an opportunity for you to counter-attack. Ideally, you will have either some air assets or artillery assigned to you to blunt the enemy thrusts. The reserve will also enable you to deceive your enemy. For example, moving them around behind your line will keep them guessing as to what your intentions are. Deception is the fifth principle of defence.

Lastly, troops do not like to defend. Defence means sitting in a position and taking what the enemy chooses to inflict upon you. It is psychologically draining. So as far as possible you must maintain an offensive spirit among your troops—this is where your reserve can be very effective in carrying out smaller counterattacks or flanking attacks on the enemy—not only to inflict damage on the adversary and complicate their decision-making, but also to keep up the morale of your troops.

If this is not possible you must be confident that your troops understand clearly why they are defending—is it to hold the enemy in this location so another counterattack can go in elsewhere? Is it to blunt the enemy assault and inflict as much damage on them as possible? Linking your troop's defence of an area to a higher purpose will make it much more effective and ensure that your soldiers are much more likely to survive.

Fighting a combined arms battle

Now we come to the nuts and bolts of offensive combined arms manoeuvre. You will recall that the centrepiece of your army is a triad between infantry, tanks (armour), and artillery. Each of these capabilities has weaknesses and strengths that, when combined, make a potent and effective fighting force that armies have used for over 100 years. Let us talk through an imagined battle as a way of demonstrating how the triad, and other capabilities like engineers, reconnaissance troops, air power, cyber, and information warfare might be used to convince your enemy that defeat or surrender are preferable to their own deaths.

Ahead of your main body of tanks and infantry (whether armoured infantry in armoured fighting vehicles, mechanised infantry in fast road vehicles, or dismounted light role infantry), you must have some sort of reconnaissance screen (a 'recce screen'). The job of the recce screen—mounted on fast, lightly protected vehicles, motorbikes, or on foot—is to probe forward and find where the enemy is, or to confirm already existing information about your opponent's locations and strengths that you already have.

In this, they will be aided by any other reconnaissance or surveillance capabilities that you can deploy—ideally drones loitering over the battlespace continuously identifying enemy targets, or satellites looking for heat signatures.

For instance, air surveillance is particularly important if your enemy is employing reverse slope defence (see Fig. 16). This is where a force deploys its units on the opposite side of a hill from the direction of their enemy threat. This protects them from direct fire—that is, tank and infantry fire—and protects them from observation, which is needed to make artillery fire accurate. It also enables them to ambush any enemy formations at short range as they come over the crest of the hill. As with intelligence, you want to have as many different methods of finding enemy combat power as possible: when combined they are extremely powerful.

There are two broad types of reconnaissance that your ground recce screen should employ: recce by movement and recce by fire. The former is what I have just described—advancing over the landscape into likely or known enemy concentrations until the enemy is found, or the enemy attacks the recce screen. However, the enemy may be particularly disciplined, and small units may remain hidden, and may not fire at your recce screen as it goes past—they know that their job is to engage your main body, and to do that they need to avoid your recce screen.

A method of getting around this is for your recce to conduct reconnaissance by fire, which involves firing into areas of likely cover—perhaps there is a thick, dense wood, or an old, abandoned factory in which might be hiding an enemy company. In this way, if there are enemy units there, they may consider themselves under attack, and return fire, thus confirming their position.

Whichever methods are employed, your reconnaissance forces must aim to find as many of the enemy as possible, to understand all you can about them (strengths, weapons systems, morale, vulnerabilities), and critically, to keep their forces away from your main body of troops (to stop them reconnoitring you). Your reconnaissance screen should seek to keep the enemy at arm's-length until you are ready to strike.

Attached to your recce you may decide to have engineer support. Your engineers need to understand the battlefield in terms of their role—mobility of your own forces, and counter-mobility of the enemy's forces—and with this in mind assess key pieces of terrain and infrastructure. Later, as your main body passes through, your engineers will build bridges, roads and defensive positions.

We are now approaching the point that the principal contingent of your troops will be nearing the enemy's main body. Exactly how you fight this battle will depend on your enemy's capabilities. For instance, if they have massed infantry, but little artillery or armour, then you may decide to stand off and destroy them from afar with artillery or by using air power. This will work less well if they are dug in trenches or bunkers, in which case you may need to fix them with artillery before assaulting them with tanks and infantry.

If both you and your enemy have lots of artillery, then the outcome will depend upon who has the artillery with the longest range and can supply ammunition more effectively and efficiently. In stark terms, if you have longer-range artillery that is well supplied with ammunition, you can keep your enemy further away from you.

Even more importantly, if you can keep their supply dumps a long way off from their front-line troops, fewer supplies will get through. If you can establish 'fire control'—that is, artillery coverage—over a main supply route, road junction, or logistics dump then it will severely reduce the volume of supplies that your enemy can distribute down that particular logistic chain.

The other key role of artillery when your enemy also has effective artillery is counter-battery fire. You must use your artillery to destroy the enemy's artillery batteries—which is obviously much easier if you have longer-range artillery than them. There are two main methods of doing this. You can launch aerial recce, particularly UAVs, to find the enemy gun positions or, if you have it available, use an artillery- and mortar-locating radar.

This ingenious piece of equipment detects projectiles in flight, instantaneously calculates their point of origin, and feeds the coordinates to your own artillery. With a highly trained crew, you can direct artillery fire onto the enemy battery within a minute or two. This forces the enemy to fire their artillery and then move to a new position immediately, thereby reducing their overall rate of fire. Almost every large coalition base in Afghanistan had one of these radars, which significantly reduced the number of accurate incoming rockets.

If your enemy has similar technology, you will be forced to do the same, which leads to a so-called artillery duel. If you find yourself in one, then your best option is to locate the enemy's counter-battery radar by searching the electromagnetic spectrum for radar emissions and then destroy it. Of course, your enemy will be trying to do this to you as well!

Once your artillery have finished duelling, you may find yourself in a situation where your tanks and infantry come face-to-face with enemy tanks and infantry. There are various possible permutations. First, the idea of fighting the enemy only using your tanks should be avoided. Tank-on-tank battles are extraordinarily rare, mostly because anti-tank missiles have a longer range (upwards of 4km) than the main armaments on most tanks (usually about 3km). This means you can set up an infantry screen with anti-tank weapons that holds the enemy armour off at a range greater than their cannons can fire.

Even if the enemy is unable completely to destroy your tanks, it is fairly easy to conduct what is known as a 'movement kill' (or M-kill) where the track mechanism is damaged and the tank effectively stuck in one place. They can then be destroyed at leisure. Alternatively, the shrapnel from airburst artillery can be used to strip all the antennas from the outside of the vehicle thereby blacking out all its communications. Either way, the tank is now a lot less useful, and can be dealt with at a time of your choosing.

Infantry also help you clear land mines, and reconnoitre key chokepoints like bridges, and assault positions (see below). You may find yourself in the rare situation where your tanks confront the enemy's, but this is extraordinarily risky because they may have infantry close by, which would put your tanks at grave risk. The general rule is that you should never risk your tanks without close infantry support. When mixed in with mechanised or armoured infantry, however, tanks are very useful for proving speed and punch to an advance or assault. Your best option is to use tanks and infantry together—the infantry protects the tanks, and the tanks can use their main armaments to inflict greater damage on enemy vehicles and fortifications. If you throw in artillery to this mix, then you will have the unbeatable triad.

If you are unable to destroy the enemy's positions with artillery—perhaps they are dug in, or you have insufficient artillery ammunition, or your guns are not of suitable calibre—and you have fought a tank battle, or there is no tank battle to fight, you now must consider how to assault the enemy's positions. The basic mechanics of the rule of three come into play—you will need troops to assault, suppress, and to maintain a reserve. This is the essence of why you need three times as many troops to attack as to defend, and why the rule of three guides not only the structure of your ground forces, but also the overall size of your force.

First, you need some troops to suppress the enemy—to fire at them to keep them occupied, with their heads down, to avoiding being killed. It stops the enemy organising a defence and returning fire onto your troops. The suppressing role can be carried out by artillery or mortars, or heavier calibre weapons and you may find it easier to place your suppressing troops on higher ground (if available) so that they can fire down onto the enemy position.

That said, no enemy worth their salt will allow unoccupied high ground next to their positions—the most sensible option is to use high ground to build defensive positions because they command the surrounding terrain. This keeps the enemy busy, giving you the freedom of movement to manoeuvre your assaulting troops into position ready for the assault.

Once you have suppressed the enemy, you can assault their position with another contingent of your troops. Now is a good time to jam their communications if you have any electronic warfare assets available. In the most basic configuration, you would have your suppressing and assaulting troops positioned at right angles to each other, so that as the latter advances, the troops who are suppressing the enemy guide their fire so that it lands just in front of their line of advance thus keeping the enemy's heads down until the last moment (see Fig. 17). Exactly which types of troops you use in the assault will depend on how the enemy is configured (and of course what you have available).

One option is tanks supported by infantry either dismounted, or in armoured personnel carriers. Which you choose will depend upon the level of anti-tank weapons held by the enemy, or the presence of minefields. Assuming the enemy lacks these capabilities, then a mechanised or armoured infantry assault will enable you to assault the enemy position with the minimum of casualties. The job of your armoured vehicles is to deliver your infantry to or near the objective.

Most likely, you will keep your tanks outside the position cutting off likely escape routes and providing over-watch, while your infantry-carrying vehicles—who have their own machine guns and small cannons to engage the enemy on the approach—will either go into the objective or stop just outside where they will deposit your infantry. Your infantry will now have to fight through the enemy position.

If the level of threat to your vehicles is extremely high then you may consider an assault conducted entirely by dismounted infantry. Dismounted infantry are, of course, vulnerable to most battlefield weapons, and so you may take many casualties assaulting in this fashion. Other options include using infantry to take out the enemy anti-tank defences, thus clearing the way for your armour to assault the position. But you may have no other choice than to use dismounted infantry.

There is no easy solution to assaulting a well-defended enemy position, and whichever option you decide upon, the assault must be conducted with the maximum possible aggression and violence. At the precise moment of the break-in to the enemy position, you are aiming to psychologically overwhelm the enemy so that for a brief moment their decision-making is overloaded and terror overcomes them. You need this moment to be long enough so that their unit cohesion breaks down and they either run away, surrender or are killed.

This is the central point of war. Your troops will have to kill enemy troops on the objective in close-quarters combat. This inevitably involves hand-to-hand combat with bayonets, grenades, rifles and pistols (before storming a position, your soldiers should fix bayonets so that if their rifle jams or they run out of ammunition they can stab the enemy in the chest). Maximum aggression is required, and your soldiers must be psyched up to kill; otherwise, they risk being killed themselves.

The brutal nature of this reality explains why infantry soldiers are the centrepiece of any army, and the storming of an objective with hand-to-hand fighting the centrepiece of combat. The unavoidable reality is that if the enemy wants to hold onto terrain—particularly if it is a village, or in a forest, or on a raised ground such as a hill, and they have had time to prepare a trench system in the ground—then the only way of dislodging them from that ground will be with an infantry attack culminating in close-quarters hand-to-hand fighting. Such infantry assaults decide who wins wars, and if one side is willing and able to conduct them and the other not—assuming they are broadly matched in other areas—only one side will prevail.

Prisoners and the rules of war

Inevitably at some point during your war you may have the opportunity to take enemy soldiers prisoner. How to surrender (raise your hands, display the white flag, communicate that you wish to surrender), and how to treat prisoners (give them food and medical attention, protect them from battle, do not subject them to degrading treatment), are governed by international agreement which is enshrined in various treaties and legal instruments that together make up the rules of war (or laws of war).

Whether you choose to abide by these conventions is of course up to you, but there are several advantages of doing so. First, the practical: taking prisoners and treating them well (and being seen to treat them well) means that enemy soldiers are more likely to surrender, which is often much easier than having to kill them. Conversely, mistreating or torturing prisoners is more likely to encourage your enemy to fight to the death, which will put more of your soldiers in harm's way and will make your task of defeating the enemy much more difficult.

In informational terms, you will lose wider consent for your war if you are known to be, for example, mistreating prisoners. The Americans and British got this badly wrong in Iraq where they allowed undisciplined troops to mistreat Iraqi prisoners, thus fatally undercutting their rationale for the war—that they were liberating the Iraqis from a tyrannical government.

Far better that you create a situation where the enemy are surrendering to you, and your own troops are treating them well (and not surrendering themselves). The wider world will draw inferences about which combatant has a higher moral purpose—soldiers tend to surrender more when they don't believe in what they are doing, or they believe that their own government has sold them out or sent them to fight an unjust war. The world, not to mention the home population, will look at the side that is surrendering en masse and identify it as a protagonist engaged in a purposeless war.

Finally, but not least, you should consider legality. Mistreating prisoners is illegal under international law, and as the leader of combatants committing war crimes you could find yourself in a courtroom answering charges (quite apart from the 'image' problem). Realistically, this is unlikely—many more people commit war crimes than are punished for them—but if you lose the war it becomes much more likely that you will end up in the International Criminal Court.

Beyond prisoners, there are wider aspects of correct, proper and legal behaviour that you should consider when fighting a war or using military force. There are four broad principles that you should consider: military necessity, distinction, proportionality, and humanity.

The principle of military necessity states that your application of military force should be directed at and assist in the defeat of enemy forces. This would mean, for example, that taking revenge on the enemy for an attack by bombing a city would be considered contrary to this principle.

Related to this is the principle of distinction between military and civilian targets. In some wars this can be quite straightforward—if they are in uniform and carrying a weapon, then they are a military target. However, it can also become very difficult—what if one of the sides has ordered a general mobilisation of the population, and ordered everyone to fight an invader? What if there is an insurgency? It could then be argued that everyone, or all men between the ages of 15 and 50, are targets. Whether or not it is difficult, you must try to distinguish between military and civilian targets.

The principle of proportionality accepts that there may be damage to civilians in the pursuit of military objectives, but states that the damage should be proportionate to the scale of the military target. Attacking a huge enemy logistics base, but in the process destroying a civilian house, might be considered more proportionate than, say, destroying a city to kill one enemy sniper.

In addition to this principle, there are certain categories of building which are always to be protected in war, including hospitals, religious and cultural sites, and nuclear power facilities. The historic Bridge of Mostar was destroyed by Croatian forces during the Balkan Wars of the 1990s. The Croatian commander claimed that it was a strategic military target used to transport supplies, but it was successfully argued in court that his forces had targeted cultural property with no military value. Beware!

Finally, the principle of humanity (sometimes called unnecessary suffering): act with humanity and do everything you can to minimise human suffering in the pursuit of your military objectives. In some cases—for example, the Chemical Weapons Convention, discussed in Chapter 8—certain categories of weapons are already banned, partly because of the suffering that they cause. Some countries, like the United States, review new weapons to judge whether they cause unnecessary suffering or not, and your soldiers should consider this when using weapons in an unintended manner—for example, dropping white phosphorous onto an enemy target rather than in front of them to screen your own troops.

It can be very difficult to apply these principles in the heat of battle, and as difficult later to prove in a courtroom that they have been contravened. (It is also not clear yet in international law whether mercenaries/private military contractors are covered by the laws of war.) Yet the commission of war crimes will cause you practical, informational and legal problems. Consider attacking civilians: this is likely to inspire fanatical resistance (they have nothing to lose), world opinion will turn against you, consent for your war will evaporate and your military commanders may well end up in court after the war. Many commanders from the Balkan wars and the continent of Africa have ended up in court and jail. It is unfortunately less likely that you will be prosecuted if you are from a powerful or victorious country.

For these reasons and moral ones, most states try their best to avoid war crimes, and so simplify the laws of war so that they are easily understood by highly stressed personnel in battle. Do not deliberately target civilians. When conducting planned attacks, carry out an assessment of estimated civilian casualties. Accept prisoners and treat them well. Don't attack schools, hospitals, water sanitation plants, or places of worship. Allow civilians to escape from sites of battles by calling ceasefires. Aim to kill your enemy rather than causing them unnecessary suffering. Most soldiers, except for the exceptionally dehumanised, know what war crimes look like. All civilians certainly do.

Some states go even further than this and issue 'rules of engagement' to their militaries. At the extreme, soldiers may only use lethal force in self defence. Or perhaps they are only allowed to target individuals when they are carrying a weapon. Attacks may only be allowed to be carried out if there is an expectation of zero civilian casualties. Hence it is advisable to issue rules of engagement to your military, backed up with disciplinary action or punishment for infractions. This will help you as a commander or leader keep control of the practical, informational, and legal issues that arise from the committing of war crimes by your troops. It's also the right thing to do.

Conclusion:
How to End a War

By now, you will have a good understanding of how to fight a war. You will know that in war psychology is paramount, and intangible factors—like strategy, logistics, and morale—are fundamental to a successful outcome.

But during any war, there will come a point where talk shifts to how the conflict should be brought to a close. Sometimes this will be early on, and these discussions will likely be driven by the sense that a terrible toll on fighters and civilians is being enacted, and that anything would be better than continuing the war. I would resist these siren calls because unfinished wars, where a 'peace' has been imposed, tend to start up again. Remember, war settles geopolitical questions that we have failed to solve in any other way.

There are two other broad scenarios in which wars end. The first is what is known as a mutually hurting stalemate. This is where both sides accept that they cannot win the war, that they won't lose it either, and the continuing fighting is damaging both sides to such a degree that they wish to stop. The impasse to be overcome in arriving at this point is persuading politicians and generals—particularly those that started the war or are intimately involved in its prosecution—to accept that they cannot prevail.

Human psychology is such that 99 per cent of leaders would continue to fight a mutual stalemate rather than accept that they were wrong in the first place. Make sure you are in the 1 per cent.

The final scenario in which wars end is where one side defeats the other's military in the field. Whether the defeated party is occupied is not important; that they can be occupied and dictated to is. One side has total power over the other—how does it use it?

This is a time where emotions will be running the highest. You have finally gained victory over the enemy who have been bombing, shooting and killing your citizens, destroying your towns and cities, trying to remove you from the pages of history. The instinct that 99 per cent of people would feel would be to enact revenge, demand reparations, and put their prisoners to work rebuilding your country.

Again, I would caution that should be in the 1 per cent. Acting with strength and magnanimity will gain you a lasting peace. Acting out revenge—no matter how 'justified'—will open another cycle of killing in the future. Humans have very strong in-group out-group boundaries, and you—at the height of your powers having just won the war—must seek to soften the boundaries between your people and theirs, which requires both strength and magnanimity.

Even more so than fighting wars successfully, ending wars conclusively requires the most renowned statesmen and women. They must have the broadest of strategic horizons capable of not only seeing clearly world geopolitics now but also envisaging how they may evolve in fifty or a hundred years' time. When I cast my eyes around the global landscape, I see no-one of sufficient clarity of thought, and greatness of vision. But, my God, do we need them.

Epilogue:

The Future of War

We live in an age of unprecedented technological advance, and this applies nowhere more than to the field of war. From hypersonic missiles and nanotechnology, through space warfare using satellite-based lasers, to biologically-enhanced soldiers, barely a week goes by without news of a new technology that will change warfare forever.

Particularly in the West, we are fascinated by how technology will give us the edge in war. Possibly because it always has. From weighted spears, to ironclads, to the hydrogen bombs and drone swarms—the West and its antecedents have almost always brought greater levels of technology to the battlefield. And for 500 years, if not longer, this has enabled the West eventually to win most wars.

Other countries see technology as the key too. In recent years, other countries—notably Russia and China—have attempted to change their militaries from manpower-intensive attritional forces to high-tech manoeuvrist forces. Judging by Russia's performance in its 2022 invasion of Ukraine, it has failed in this endeavour. It is not yet clear whether China has been able, or will be able, to make that transition.

Yet, the main argument of this book is that war has a series of principles that you cannot wish away—space-age technology will not compensate for a lack of a strategy, nor the poor morale of your troops, nor non-functioning logistics.

Warfare will continue to be about people, and a phenomenon rooted in human psychology. Wars will still be won when one side decides that they have had enough. Wars will still be won with better strategies. And strategies will still consist of the same dynamics of advancing, retreating, deceiving and creating fear among your enemies. War will remain predominantly a human emotional activity, rather than a technological one.

One of the fallacies that we encountered in the Introduction is that technology will change warfare. It is a classic mistake that leaders make when embarking upon wars. They believe that one special technology will allow them to ignore the basic principles that form the basis for this book. But, as we have repeatedly said, technology will not change the nature of war. Or will it?

Well, there may be one scientific development on the horizon that will change warfare forever. That technology is artificial intelligence, or AI.

Humans will no longer be making decisions. Artificial brains will. Whereas human brains win wars using methods with which we are so familiar—bluff, advance, entrap, deceive—this may not be how AI does so.

Those methods are a product of brains that evolved in a specific evolutionary environment, in social competition with other humans. We all share the same emotional responses (more or less). We are all jealous, angry, proud, and sad. These emotions, and these ways of psychologically relating to others, are the basis of strategy. What is a ruse, after all, if it is not playing on the opposing commander's pride?

We know little about the future AI systems that will be running wars, except that, in this most competitive human field, there definitely will be AI systems doing so, and that they will not look or act anything like our human brains. AI systems will think differently. The central psychology of war will disappear.

Human brains that evolved through competition with other human brains create the psychology that underpins strategy, which creates the essence of warfare. But these human brains evolved in order to maximise survival and reproduction: to find food and water, and sexual partners, to form coalitions, and to avoid being killed by lions.

Artificially intelligent war brains will have none of these ultimate goals. AI's only goal will be to win the war (one assumes it will be programmed accordingly), and so the strategy it adopts will have an entirely different shape. That's why, if applied to the battlefield, AI could change the essence of warfare in a way that is beyond human speculation. Artificial intelligence is likely to be the single most important technological innovation in the field of conflict, ever.

Thus, for the first time in human civilisation, the fundamental nature of warfare may change. And it will look completely different, with different dynamics, in ways that we cannot even conceive. In short, AI systems will create the new nature of war.

My argument relies on AI making the strategic decisions, and at present (2023) we are unlikely to see this for some time. But automation is already being introduced at lower levels. Advanced militaries are already designing and testing autonomous weapons systems like loitering missiles and drones (the US, for instance, spends about $2bn per year on researching autonomous systems). Why? Autonomous systems are much, much faster than humans at making decisions, they don't get tired and need to sleep, and nor do they become casualty statistics. The press has dubbed these 'killer robots' and articles published on the topic are invariably illustrated with a still from the Terminator movies series.

If I were a general or a leader of a country—that is, if I were you—I would be observing these experiments very closely. In a few years (or maybe even already, but still unknown to us) autonomous systems will start to fight each other. And the dynamics resulting from that encounter will reveal what the future of war will look like.

INDEX

INDEX

INDEX

INDEX

INDEX